P9-CEL-410

POWER BEHIND THE WHEEL

POWER BEHIND THE WHEEL

CREATIVITY AND THE EVOLUTION OF THE AUTOMOBILE

WALTER J. BOYNE

COLOR PHOTOGRAPHS BY LUCINDA LEWIS

Stewart, Tabori & Chang

NEW YORK

Published in 1988 by Stewart, Tabori & Chang, Inc.
740 Broadway, New York, New York 10003

Library of Congress
Cataloging-in-Publication Data

Boyne, Walter J., 1929–
Power behind the wheel : creativity and the evolution of the automobile / Walter J. Boyne ; color photographs by Lucinda Lewis.
p. cm.
Includes index.
ISBN 1-55670-042-3 : $35.00
1. Automobiles. 2. Automobiles—Technological innovations. I. Lewis, Lucinda. II. Title.
TL154.B653 1988
629.2'31'09—dc19 88-15289
CIP

Distributed in the U.S. by Workman Publishing, 708 Broadway, New York, New York 10003
Distributed in Canada by Canadian Manda Group, P.O. Box 920 Station U, Toronto, Ontario M8Z 5P9
Distributed in all other territories by Little, Brown and Company, International Division, 34 Beacon Street, Boston, Massachusetts 02108

Printed in Japan

10 9 8 7 6 5 4 3 2

(Photograph by John Vachon)

Acknowledgments

In this, my first book outside of the field of aviation, I feel deeply indebted to a host of people. First and foremost are those men and women who have written the great books and inspired the great magazines, from Beverly Rae Kimes and Henry Austin Clark, Jr., to Griffith Borgeson and Eugene Jaderquist, to the late Michael Sedgwick and G. N. Georgano. There are so many others!

I was privileged to work with Cindy Lewis, who is as remarkable a photographer as she is a businesswoman, pilot, entrepreneur, and human being, and whose photos bring marvelous color to these pages.

Similarly, I had wonderful support from the entire staff at Stewart, Tabori & Chang. Andy Stewart always has time to listen, Roy Finamore is a sensitive but demanding editor, and Sarah Longacre headed a picture research staff that included David Diamond and Jose Pouso.

And as a final note, I want to tip a grateful automotive hat to my literary agent, Jacques de Spoelberch, who encouraged me in this as tactfully and as beneficially as he has in all our ventures.

Walter J. Boyne
Reston, Virginia
May 21, 1988

DESIGN
Paul Zakris

PHOTO EDITOR
Sarah Longacre
ASSISTANT
David Diamond

Composed in Futura
by Arkotype Inc.
Printed and bound
by Toppan Printing Company, Ltd.

CONTENTS

After years of producing the epitome in conservative, luxurious styling, the venerable Pierce-Arrow company startled the world by presenting its Silver Arrow at the 1933 New York automobile show. Its sleek envelope body forecast the future while its silk-smooth V-12 engine met all the contemporary demands for multiple-cylinder power.

INTRODUCTION

The automobile, the ultimate mechanical love object, has been with us for just over 100 years. In that time it has served as both the symbol and the instrument of some of the most profound changes in history. Humanity recognized in the automobile a signal to launch a period of unprecedented expansion, prosperity, individual fulfillment, aggression, pollution, and distress. The vehicle changed dramatically over time—from tiller-controlled carriages to chauffeur-driven Mercedes, and from Model T's to Lamborghinis—but no matter what their type and their degree of engineering excellence, all inspired loyalty and unabashed affection as they altered the world.

The Age of the Automobile began rather simply, building bit by bit on the inquisitive tinkering of such men as Karl Benz, Gottlieb Daimler, Henry Ford, and Charles Duryea, who thought they were fashioning an alternative to the horse when in fact they were standing at the top of a massive snow mountain, patting together a rolling shape. When this coughing, sputtering, mechanized snowball began its rumble into history, it picked up the momentum of a juggernaut, absorbing the giant personalities of all who had a hand in conceiving and furthering it: the Daimlers and Fords, and later the DeLoreans and Iacoccas, too. It grew into a gigantic force that dictated economies, won wars, depleted the resources of 600 million years, and spread wealth farther and more evenly than economic philosophers of an earlier age could possibly have conceived.

OPPOSITE:
Whether it graced the running gear of a Rolls-Royce or a Ford V-8, the famous American coachmaker Brewster's distinctive motif—heart-shaped grille and flared front fender—was highly prized by discriminating owners.
BELOW:
The Citroën 2CV was the car thoughtful people demanded: simple, economical, disdaining inane styling trends. Unlike most cars that also met these standards, however, it was popular as well.

A curious aspect of the automobile has been the manner in which men and women of all countries view it—not in terms of need, but of desire. With notable exceptions—the Ford Model T, the Renault 2CV, the Austin 7, the Fiat Topolino, and the early Volkswagen, among others—cars escaped the utilitarian mode, becoming instead a statement about the owners themselves. The statements varied: some people aspired to the aristocratic hauteur of a Rolls, while others longed for the gutty sexual elegance of a Ferrari Testa Rossa, and still others sought the simple respectability of a Buick. There were even role reversal statements: Howard Hughes drove plain-jane Chevrolets, and what in another generation would be called yuppies drove VW Bugs. Derivatives of these desires permeated the serried ranks of the economic pecking order. Englishmen who knew they'd never have the money for a Rolls instead bought a miniaturized version, the Triumph Mayflower, not caring that the crisp Rolls lines looked absurd on a 90-inch wheelbase. Their American counterparts, thirsty for distinction, purchased the vastly underrated Crosley Hotshot—not for the excellent perfor-

mance the car delivered, but in the vain hope that it would be mistaken for an XK-120 in someone's rearview mirror. Would-be Buick owners often sank bolt-on portholes into their Chevrolet fenders.

Of the same ilk are the countersnobs who claim to disdain automobiles and insist upon driving rusty, wheezing Volvos for years after the monthly repair bills exceed comparable new car payments. These car haters are in fact closet classic car lovers who, frustrated because they can't have an Isotta-Fraschini or a Duesenberg, stick to their ancient rigs in order to call attention to themselves.

Remarkably, the car's introduction preceded the airplane's by only eighteen years. Part of the pleasure—and part of the pain—in charting the motorcar's progress relates to the inextricable intertwining of these two engineering processes. Less obvious is the incredibly strong psychological connections people established between the two forms of transportation—connections characterized by strong similarities and even stronger dissimilarities.

In chronological terms, aircraft engineering followed a longer and more sequential path. The Wright brothers' aircraft, sensational though it was in 1903, was barely capable of an obviously impractical flight regime. The Kitty Hawk Flyer was, as Benjamin Franklin once observed of the balloon, as useful as a newborn baby. Its practical development would be protracted; thirty-three years of hesitant progress elapsed before the creation of an airplane (the Douglas DC-3) that could generate a profit for its operators independent of government subsidies. The next fifty years saw amazing improvements in aircraft performance, nurtured by military expenditure and commercial profitability. Progress in the aircraft industry was built upon successive steps of engineering development, each one representing a leading edge of technological capability.

The automobile matured much more rapidly than the airplane, its commercial success founded not at once upon utility but upon pride of ownership and the personal freedom that four wheels and an engine imparted. Perceived as the key to individual liberty, carrying its owner through the

ABOVE:
Mobility for the family: the photographer's shadow perhaps bears best witness to the owner's pride of possession in his Model T and in his family. *(Photograph by Otto M. Jones)*
OPPOSITE:
Mercedes made more than a virtue out of a necessity with the stunning gull-wing doors of the 300SL. The tubular construction of the chassis required an innovative means of entrance, and the solution bestowed instant status as a classic on the car.

city and along country roads, the car achieved a universal desirability unmatched by any other mechanical contrivance. The automobile's swifter progress to economic success stemmed from the fact that its engineering challenge was easier to master than the airplane's. A skilled pilot was needed for an aircraft, which had to be controlled to the left and right, fore and aft, *and* up and down! The driver of a car had only to contend with its direction of travel, and with turning. The car was probably easier to learn to drive than a horse-drawn carriage. And then there was the matter of safety. If an automobile engine quit, the car would stop, sometimes inconveniently, to be sure—in traffic, or along an isolated country road—but normally without imperiling the driver. When an aircraft engine stopped, it was a matter of very different import. The airplane began to descend immediately, and the pilot had to be fortunate indeed to be situated at that moment over a field in which a landing could be made.

More pertinent, however, is the fact that the normal performance parameters of the automobile were defined and achieved very early on. The car owner's genuine need for speed is limited by the operating environment; to go from home to the grocery doesn't require breaking the sound barrier. In the American West, a top speed of 100 mph permitted by open spaces and relatively good roads (although prohibited by law) was adequate for the sane. And even the wildest would-be Mario Andretti, helmeted and strapped in the hottest Porsche available, did not *need* to exceed 160 mph on the most permissive autobahn. In practical reality, maximum speeds of 45 miles per hour around town and 65 on the highway were entirely satisfactory for most. The need for speed and the desire for it are irreconcilable, however, as casual perusal of any monthly car magazine will prove; and this factor is just one element of the psychological barriers we've built to preserve us from true "people's cars."

In contrast, each time someone extended flight performance capabilities further—whether it was Howard Hughes in his sleek silver and blue racer,

The word *soigné* perhaps best categorizes this raccoon-coated pair from the 1930s haut-monde in their lovely Cadillac V-16. *(Photograph by James Van Der Zee)*

or Dick Rutan and Jeana Yeager in their ethereal white Voyager—new paths were blazed. More sophisticated airfoils were created, weight was reduced, more power was squeezed from new engines. The refinement process always seemed to grasp past practice by the throat, shaking it to gain new goals. The reason lay in flight's central equation, wherein four contending forces—lift, weight, drag, and thrust—must always be balanced to achieve the new performance equilibrium. These four forces are not equally easy to change: weight and drag are always easier to increase than lift and thrust. The four forces have their automotive counterparts, but with the essential and obvious difference that the aircraft flies through thin air, while the automobile has the less difficult task of rolling across the ground.

There are other fundamental reasons why the automobile had an easier path of development. The weight of an aircraft has always been critical; the Wright brothers' plane, for example, was just light enough to fly. In contrast, it was almost irrelevant what the first cars weighed. Today, more than eight decades later, the Rockwell International B-1B bomber is assailed by critics as being too heavy, and the same charge has been leveled at almost every aircraft ever built; modern plane manufacturers have assigned enormous numbers

of engineers to the weight reduction process, where they dedicate themselves to shaving not only pounds but grams from individual components. In contrast, heavier weight in cars was for years taken as a measure of status and esteem.

In reality, of course, excess weight hinders an automobile's performance, but it rarely prevents it from rolling. There are currently many advocates of the heavy motorcar, extolling weight for safety's sake; and each year when the statistics are released, it is evident that people are safer in a Mercury Grand Marquis station wagon than in a tiny imported econobox.

Thus, while the Wrights and their colleagues had to agonize over how many millimeters thick a piece of fine-grained wood could be, the early car makers could adopt standard carriage practice and incorporate hefty wood-and-metal structures that easily absorbed the power of the engines and the stress of driving over bad roads.

Learning to control the flight path of an aircraft was far more difficult than learning to steer a car. Of all the early aviation pioneers, only the Wrights saw that flying was going to be essentially unstable, and that control had to be exercised around the three axes of flight. The others, who assumed that flying would be like steering a rowboat or handling a horse-drawn carriage,

Not at all the toy it appears, the Austin 7 debuted in 1922 and remained in production until 1939, actually growing more popular with time. It is graced here by Colleen Moore, star of *Lilac Time* and epic films of the period.

were terribly wrong. The early designers of the automobile thought that driving would resemble operating these existing forms of transportation, too—and they were right. Steering, although continually improved through the years, was never an imposing engineering problem.

Controlling the speed of the automobile was also easier. Speeds ranged from an absolute standstill to slower than a person's walk to as fast as the motorist dared—perhaps 40 miles per hour. In contrast, sustained high speed was essential to an aircraft's flight; the wing had to be moved through the air so that lift could be created and the bounds of gravity broken. Almost every improvement to an aircraft increased its weight, which in turn meant that it had to go faster to fly. Whereas the Wright Flyer became airborne at perhaps 20 miles per hour, a jumbo jet must reach 160 mph before it can leave the ground. The difference in mass is even more incredible: the Wright's airplane weighed 735 pounds, and at its putative 30 miles per hour speed represented an energy of 21,050 foot-pounds; the 747, at 160 mph and 700,000 pounds, represents 112,000,000 foot-pounds of energy.

And if hurling loads like this into the air is difficult, safely stopping them is extraordinarily tricky. Stopping a car was obviously less of a problem. Brakes were easily made and soon became sophisticated, but a car could be brought to a halt by steering it into the side of the road if all else failed. Stopping even a lesser aircraft than a 747 required time, room, judgment, and tremendous mechanical refinement, to compensate for both ever-increasing elements of the speed × mass equation.

Sequential leading edges—devices or inventions that extend the performance of a vehicle to its farthest limits—have been the building blocks of improved aircraft performance. In contrast, progress in the automobile industry has benefited from a "rolling edge" of gradual development. Almost every element of the modern automobile, with the single exception of the computer, appeared within the first thirty years of the great machine's inception. Many features, such as four-wheel drive and automatic transmissions, were primitive and premature when they first appeared. Yet as the need arose, each was developed to a point where it could be incorporated sensibly. Over the century of automotive history, the wheel has been reinvented repeatedly, as features that were ahead of their time have been reintroduced at more propitious moments. These are usually improved, to be sure, and are often cast in a totally different light by other developments, but they retain essentially the same concept as had appeared in the first three decades.

This difference in technical progress—leading edges versus rolling edges—has been obscured by two factors. First, most people want a car to be more than it needs to be; and second, there is a natural inclination to confuse the engineering of automobiles with the engineering of aircraft—given that both are popular means of transportation, both have taken advantage of similar developments, and both are often products of the same individuals or corporations.

A single component of the automobile, the engine, illustrates the rolling edge of automotive power very well. All car developers saw the advantage of more power and immediately began approaching the problem from a number of different directions. The power capacity of the steam engine was already well known, and this provided the Stanley brothers and their colleagues with a direction. For like reasons, others chose electrical power as the energy source to develop in cars. Still others chose to stay with the internal combustion engine and elected alternatively or in combination to add cylinders, change the configuration from in-line to V-shaped to flat opposed, increase compression, provide more valves per cylinder for better "breathing," add supercharging, and improve carburetion—all without an aeronautical degree of risk. Periodically, a radical departure would occur: six-stroke engines, rotary engines, radial engines. The list was richly varied. Every one of these features has been brought forward as an innovation repeatedly over the years;

PAGES 14–15:
This fabulously beautiful 1933 Rolls-Royce Phantom II was the result of a little competitive one-upmanship: Film star Constance Bennett purchased a very elegant Rolls with a cane-sided body. Not to be outdone, the Countess di Frasso, Bennett's friendly competitor in parties and in cars, commissioned Dutch Darrin to design this stunning aluminum-bodied response.
PAGES 16–17:
In 1951, owning a red M.G. was about as close as a young couple could come to definitive motoring happiness—roughly the same as today's yuppies derive from their BMW 735i. Uncomfortable in bad weather and not at all fast, the M.G. nonetheless had a tremendous effect on the American car-buying population. ("American Couple in Red M.G.," 1951, © *Ruth Orkin 1981*)

many of them are being advertised as such today.

But any discussion of automobiles and aircraft in history must deal in the larger context of social and economic values, and for this we need some definitions. A *motor* imparts motion to perform a function (as does, for example, the motor in a vacuum cleaner or in an electric razor), but it receives its energy from an outside source. An *engine* also imparts motion to perform a function (as does the engine of a railroad train or of a gasoline-fueled lawn mower), but it generates the power itself, internally.

Rugged roads and unsophisticated suspension systems readily conspired to make automotive outings into less-than-soothing events, which were often complicated by unpredictable obstacles. *(Photograph by Ernst Haas)*

From these two definitions, we can draw an engineering analogy to describe the difference in the effects of automobiles and of aircraft upon society and the economy. For more than half of its life, the aircraft was an economic motor, driven by the power of the general economy for military and experimental reasons. In contrast, the automobile almost immediately became a great economic engine, driving the general economy of the world in ways that would have been impossible to imagine a century before. The aircraft was carried by two wars and by doting governments for

Laurence Olivier, Merle Oberon, and a Lagonda Rapide.
OPPOSITE:
A 1936 Auburn supercharged, eight-cylinder cabriolet *(top)* provides the foil for Errol Flynn, while Marguerite Clark and Mickey Neilan's chosen chariot *(bottom)* was a 1918 Marmon 34.

almost half a century before it began to establish itself as a cause rather than an effect, and even then it gripped the imagination of only a small (though devoted) portion of the public.

In striking contrast, the automobile was from its first decade a powerful economic force; and by its third decade, it had laid claim not only to the heart but to the soul of the western world. As time passed, emerging economies embraced the prospect of four-wheel freedom with consistent eagerness, discounting the smog, the accidents, and the debt that were unavoidable concomitants of that freedom. The lust for personal transportation lurked just behind hunger as a human priority.

The sequence has been similar everywhere in the twentieth century. With the first glimmering of prosperity anywhere in the world—South America, Asia, Africa—came wheels of some sort. For heads of state it took the form of stately black limousines, while for the man in the street it could begin with a bicycle. The steps of any mobility revolution were outlined in detail after World War II. Prior to that

SUPER-CHARGED

great tragedy, cars had been exclusive possessions of the well-to-do in every country but the United States. After the war, however, the masses were able to get their hands on transportation. It was often primitive: two-cycle scooters encumbered the streets of Rome, three-wheeled motorized jinrikishas jostled tourists in Hong Kong, and dazzlingly ornamented jitneys with passengers hanging on the sides sped around Manila. But the actual vehicle didn't matter, as long as it provided that intoxicating feeling of liberty on wheels. Behind these rudimentary harbingers of individual freedom would come real cars—old American, or new Asian—cars with insistent horns and blaring radios and every other accoutrement, answering this universal twentieth-century urge to roll unceasingly from point to point.

Such intensely personal, emotional, and economic clout made the automobile far more important as a social and cultural determinant than the airplane. Travel by air was exciting and glamorous but beyond the aspirations of most ordinary people. Travel by car was the accessible symbol of emancipation, mobility, and prestige. And though the promised freedom might be illusory, the desire for such travel pulses through the breast of all humanity as soon as the minimum economic threshold for attaining it is reached. Some political philosophers maintain in all seriousness that the United States has spent its money on the wrong weapons for the past thirty years. If in the 1950s the Strategic Air Command had made preemptive strikes

The automobile became standard equipment for the American family in relatively little time. Even during the 1930s, a car was a must and the ownership of motorized transportation cut across all economic and social boundaries. ***(Photograph by Russell Lee)***

on Russia and China, dropping not bombs but automobile factories, we would have created an unalterable basis for peace. Communism would have wilted as hearty muzhiks built filling stations and staunch Maoists rushed to open Pep Boys franchised auto parts stores.

In the ensuing chapters we will examine various facets of the automobile's development, from powerplants through styling, all of them elicited from enthusiastic engineers by a demanding public. The fact that the greatest advances in automotive engineering appeared early on and were often repeated is only a small part of the story. The adventure and the glamour lie in how these advances were assembled and refined into machines that captured people's imagination and often expressed both their personal and their national needs. We will also look into the ultimate irony: when the major sociopolitical issues have been solved, when hunger is a thing of the past, when everyone in the world is materially ready to drive a dream car, we will have depleted the world's supply of conventional fossil fuels. Will this mark the end of our urgent need to roll? Of course not, but it will mean that the thinking developed over the first 150 years of the Age of the Automobile will have to be radically revised to permit the use of alternative powerplants, relying on electricity generated by solar and other means, on methane gas, or (best of all) on human muscle power.

By the 1950s, cars had helped change the face of recreation for millions of Americans. *(Photograph by Robert Frank)*

OVERLEAF:

Compared to the water-cooled engines of the past, that in Lindbergh's *Spirit of St. Louis* was an air-cooled radial, bestowing unprecedented performance and reliability. The technology provided equally fine results for the Franklin automobile, but public demand dictated that a false "radiator" be adopted as well.

600-00

CHAPTER 1

A BRIEF HISTORY

A Horse, a Horse

To put the automobile in perspective, we have to examine what it replaced: the horse. And we must consider the horse of the pre-automotive situation, rather than the horse of today. This is not because of any great physical differences—if anything, today's horse is better nourished, groomed, and housed than his great-great-great grandsire—but because of the circumstances in which the horses of that time lived and worked.

In the latter part of the nineteenth century, when (as we shall see) Daimler and Benz began their important work, horses lived in close proximity to people—perhaps not right in the house, but often within 50 feet of it, and in any case not on romantic 10-acre ranchettes where they are not constantly before our eyes, in our ears, and under our noses. And contrary to the hundreds of film westerns in which we see hardy horses leaping off cliffs, dashing across prairies, and tapping out messages with their hooves to deceive the Indians, horses tend to be both delicate in health and limited in intellect. The daunting array of do-it-yourself horse medicine at a modern tack shop gives some indication that a horse is no easy possession to have around. And in view of the rigors of administering worm paste, filing hooves, and participating in certain other hygienic measures that couldn't be shown even on the six o'clock news, it's evident that a horse is a gargantuan handful.

Horses also posed genuine health hazards: each day in New York City of the ''Gay Nineties'' they deposited millions of pounds of solid waste and 60,000 gallons of urine on city streets. Nor was this the end of it. In the ordinary course of a horse-drawn carriage drive, travelers would encounter emaciated animals, horsebeating, and dead horses. Now a dead horse is not something left as roadside litter for the crows to dispose of,

The direct ancestor of all the wonderful Mercedes-Benz automobiles on the road today can be traced to Gottlieb Daimler's ''single track,'' the world's first motorcycle. Created of wood and iron in 1885, this early ''hog'' was powered by an air-cooled, one-half-horsepower engine.

OPPOSITE:
The spindly shanks of a 1900 Packard Runabout wheel give no hint of the solid future of the car. When new, the car was priced at $1,200 and worth every cent but, even though its literature stated that ''this machine has been developed with the thought in mind that it will give perfect service upon ordinary American roads,'' only five were built in 1900. This is the only known example.

UNIVERSAL TIRE CO.

just as a sick horse is not something you cuddle on your lap while you wait in the vet's office. A dead horse presents a significant logistical problem, and the automobile mercifully moved us to a new era of more tractable breakdowns.

One final note on the passing of horses is relevant. At the turn of the nineteenth century, the population of the United States was just nudging the 90,000,000 mark, and there were more than 25,000,000 horses. Imagine the conditions that would prevail today if Americans still depended on horses at the same ratio to serve their current population of 230,000,000!

Salvation from this surfeit of horseflesh came through hundreds of separate events over the last 2,000 years, which independently and simultaneously came to fruition in the latter part of the nineteenth century. The pertinent industrial factors included an increasingly sophisticated level of manufacture, the maturing of the steam engine, improvements in metallurgy, and the more intensive utilization of petroleum resources. All of these combined with extremely important social changes, including a somewhat more humane exploitation of labor in recognition of the fact that individuals were actually more productive when not forced to endure a 6-day, 72-hour work week at starvation wages, as well as the growth of a middle class that had both leisure time and some small funds to spend upon it.

The First Rolling Edges

The need for individual personal transportation independent of the horse had been felt for almost 300 years before the Age of the Automobile began in earnest. Maurice of Nassau, Prince of Orange, Marshal of the Netherlands, and Admiral of the Ocean Sea, was the Caspar Weinberger of his time. Among his innovations in weaponry and armaments was the sail wagon, an armed vehicle that carried cannon and was powered by sails. His R&D expert, Simon Stevin, constructed a huge land-based frigate, which proved capable of car-

A great hulking Peugeot hauling members of the soldiery in France, circa 1918.

999

rying a number of men at a speed averaging 25 miles per hour! Not bad for 1600.

In France, circa 1767, Nicolas Cugnot built a three-wheel steam-powered carriage, intended to replace the horses that hauled cannon about. It was a malevolent-looking vehicle that puffed its way into history at 2½ miles per hour. It, or perhaps a second version of it, can still be seen in a museum in Paris. There followed a whole series of steam engine applications to public transport that worked very well, including the notable efforts of Goldsworthy Gurney, who covered more than 5,633 km/3,500 miles operating a steam coach route around London and to points as far away as Bath after 1828.

The British government reacted with suspicion to this mechanical innovation, however, and a series of laws passed in ensuing years virtually halted its development. Among these was the Red Flag Law, which required that a man precede a horseless carriage, carrying a red flag by day and a red lantern by night. Fortunately, the law was repealed in 1896.

Nor was the desire to roam confined to the old world. Nathan Read of Massachusetts received a patent for steam engine improvements to be applied to a "land carriage" in 1790. The patent was signed by George Washington and Thomas Jefferson. Washington, a con man's ideal patsy, was always ready to invest in new inventions and was undoubtedly impressed by the idea; Jefferson, an inventor himself, may have had a more realistic view. Read was preceded in the colonies by Oliver Evans of Delaware, who had been toying with steam applications since 1772 and was the first American to pursue the idea of a steam-powered carriage. As all inventors must be, Evans was a man of considerable tenacity; he used calculations for travel over the Lancaster Turnpike, which connected Philadelphia and Columbia, Pennsylvania, as a means to trumpet the advantages of the steam-propelled wagon over its horse-drawn equivalent. He could show convincingly on paper that his mechanical vehicle was vastly more efficient, but he could not obtain investors. In 1805, he demonstrated a steam-driven mechanical marvel, the Orukter Amphibolos, an amphibious digger that would move itself over land, float in water, and dredge in rivers. The demonstration was a success, but funds were still not forthcoming, so Evans licensed his very successful steam engine for stationary use in mills.

OPPOSITE:
Henry Ford knew that he could gain national acclaim by introducing a powerful racing car, so he built two and called one the "999" *(top)*, after the record-breaking New York Central Railroad train, and the other the "Arrow" *(bottom)*. Barney Oldfield won the 1902 Manufacturer's Challenge Cup at Grosse Pointe, Michigan, driving one of these rugged vehicles at almost sixty miles an hour, thereby garnering newspaper headlines across the country.

BELOW:
A study of the wheel of a 1927 Lincoln Coaching Brougham.

The success of the steam engine for railroads was undeniable, and it led to a number of road vehicles that were used very successfully by metropolitan fire departments as early as 1855. Sylvester Roper built a formidable steam carriage around 1860 and another in 1863 that might pass for a truncated curved-dash Oldsmobile. The latter may still be seen in the Henry Ford Museum in Dearborn, Michigan. The March 14, 1863, issue of *Scientific American* described the first of the ten vehicles that Roper built as a two-passenger four-wheeler, powered by a steam engine capable of generating 2 horsepower and speeds up to 25 mph. Roper would have a long and eminently successful career in the lonely trade of inventing steam vehicles, ultimately losing his life at the (literally) breakneck speed of 30 mph in a two-wheeler in 1896—at the age of 73.

These ideas, laudable though they may have been, were premature. Roads, metals, and psyches were not yet ready, and the steam engine needed further development before it could leave the factories and rails for the road. Its successor, the internal combustion engine, had to evolve as the steam engine had, through initial stationary applications.

Who Was First?

Disputes arise about who was first with any invention. Just as the Wright brothers had to contend with the claims of Gustave Whitehead and others as to who made the first flight, the primacy of Gottlieb Daimler and Karl Benz with respect to the automobile was contested by Siegfried Marcus of Austria. And just as with the Wrights and Whitehead, the real answer doesn't matter, for the line of development of the modern automobile can

In what must have been an extraordinarily satisfying moment, one of the founding fathers of the automobile industry, Karl Benz (in 1925) driving the 1886 car that started it all. Mrs. Benz may be seen directly behind the steering tiller.
OVERLEAF:
The prewar Mercedes automobiles combined size, power, and rugged styling in a manner that was both peculiarly Germanic and extraordinarily attractive.

be traced quite clearly to the 1885 Benz and the 1886 Daimler.

Marcus's achievements are more solidly documented than those of Whitehead, however. He publicly demonstrated what can only be called automobiles in 1864 and 1865. After a period during which he was occupied with earning a living by selling other inventions to the military, Marcus reappeared with a *Strassenwagen* (street wagon) in 1874. Despite piquing considerably greater interest from moneyed backers than he had been able to achieve in the past, Marcus was unable to light the fire of economic interest that welcomed Benz and Daimler.

No one can dispute the far-reaching effects of the manufacturing and business skills of Nikolaus Otto, a hard-working entrepreneur who must be considered almost coequal with Daimler. His first firm, Otto and Langen, concentrated on developing an internal combustion substitute for steam-powered stationary engines; it began producing free-piston engines in 1867. Otto was every bit as good a personnel director as he was an inventor, for he brought Daimler in as technical director in 1873. Daimler in turn hired Wilhelm Maybach as his chief draftsman, and the critical mass for inventing the direct progenitor of the automobile had been formed. This new team conceived and brought into production in 1876 the Otto four-stroke silent engine. The new engine met with instant success, selling more than 50,000 improved units over the next twenty years. It was developed widely around the world under the basic Otto patent, which was at first a deterrent to Karl Benz.

Benz, the son of a locomotive driver, was born in Germany, in 1844, and attended the Karlsruhe Polytechnic. After bumping around in the rail industry for a few years, he founded his own firm in 1872, where he built two-stroke engines to avoid infringing on Otto's patent. The discovery that work of Alphonse Beau de Rochas invalidated Otto's patent freed Benz to develop his own four-stroke engine in a new firm; it, like the firm of Otto and Langen, was primarily concerned with production of stationary engines for industrial use. Benz, how-

ever, wanted to build self-propelled vehicles, and he created a water-cooled, one-cylinder, four-stroke engine for that purpose.

In the manner of all great inventors, he borrowed heavily from current technology, particularly that of the bicycle, to create a three-wheel carriage steered by a tiller. In similarly traditional fashion, his preoccupation with the novelty estranged him from his partners, who felt he was wasting his time and their money.

CROSS-FERTILIZATION BEGINS

Gottlieb Daimler was born at Schorndorf, Germany, in 1834. After attending school in Stuttgart, he worked in England and Germany. In 1882, Daimler and his trusted friend Maybach set up their own factory and began a process of cross-fertilization and of round-robin personnel migration. Maybach later went on to found his own firm, which built the engines that powered the feared Zeppelins of World War I. After the war, the firm entered the luxury car market with shapely, high-powered vehicles, culminating in the late 1930s in a huge twelve-cylinder car of Goeringesque proportions. Similarly, Benz left his own firm to join Panhard-Levassor. The process would continue in later years with Walter Chrysler revitaliz-

The Panhard-Levassor of 1892 set the configuration for the motorcar for the next seventy years, placing the engine in front with power transmitted to the rear wheels by a gear arrangement. This model, circa 1895, had an enormous twelve-horsepower engine lurking under the hood.

ing Maxwell and with Ford's Lee Iacocca saving Chrysler.

The first of hundreds of innovations the Daimler firm fathered was in the single-cylinder engine that Gottlieb Daimler installed in a standard horse-drawn carriage purchased from a local coachbuilder. While Benz and most other practitioners were satisfied with a very low number of rpm (revolutions per minute), Daimler souped up his engine to deliver a maximum of 750 rpm, ''cruising'' at 600 rpm and delivering a mighty 1.1 horsepower. This 10 percent increase over the power of a hay-fueled dobbin contributed to a road speed of 10 mph—revolutionary for the time.

Count Albert De Dion—whose name lives on as a term for a special type of suspension—driving his original car into a museum, a tribute most car manufacturers did not get to enjoy.

Competition in America

The Age of Steam in America had been augmented by a derivative internal combustion engine patented by George Brayton in 1872 that received wide attention and was used experimentally in street cars. The Brayton engine did not compress its explosive mixture of the new kerosene distillate, gasoline, but instead used the explosion of an air/gasoline mixture to drive the single-action piston smoothly. George B. Selden patented a horseless carriage using the Brayton engine, and thereby set in motion one of the larger patent fights of the era—one ultimately won by Henry Ford.

The 1917 Model T runabout reflected the demands of war when the brass finish of the radiator shell and decorative trim were replaced by black paint. The hoodline was raised, however, and at $345 (delivered), it was still a bargain.

The progress in all the things necessary to create a horseless carriage (improvements in the metals, the development of gasoline, and advances in the understanding of thermodynamics) was so thoroughgoing that by 1895 there was already a magazine entitled *The Horseless Age* in publication. The technology required to create a horseless carriage was within the reach of many, and all over the United States hundreds of inventors were working on the problem, any one of whom might have been the "first" inventor of the automobile in the United States.

Credit for that distinction is usually given to the brothers Charles and Frank Duryea, who later engaged in a squabble with the Smithsonian Institution on this very question that paralleled the Smithsonian's later argument with the Wright brothers over the Langley Aerodrome. Charles Duryea, who had driven a car in 1893, found an 1894 car by Elwood Haynes displayed at the Smithsonian with the legend, "First American Gasoline-Powered Vehicle." Duryea made his point and got his own vehicle exhibited as the first. Experimentation was widespread and often unreported, however, so it is highly likely that his car, too, had predecessors, including the Lambert and the Nadig of 1891 and the Schloemer of 1892.

But as with Benz and Daimler, this argument is academic. The Duryea Motor Wagon Company was founded in 1895, and the first Duryeas were just that: wagons with motors installed. The company's importance lies in its generation of the most essential of all attributes in an industry, sales. All of the hundreds of other potential claimants to be "America's First Automobile Builder" lack credibility because their efforts didn't result in the establishment of viable companies that could sell cars and inspire others to do the same.

The Lust for Wheels Spreads Worldwide

The nineteenth century was one of glory for France, whether because or in spite of the turmoil caused by the two Napoleons. For years Europe's center of culture, France was a center for technical advancement as well. When Karl Benz showed his car at the Paris Exhibition of 1887, it excited Emile Rogers, who not only bought a car but undertook to manufacture and sell them in France. In the same year, Daimler and Maybach set the pattern for much of the future with a V-2, which was licensed for manufacture by the same firm (Panhard-Levassor) that Benz would join.

The wild fertility of the early inventors' imaginations was amazing. The first Panhard et Levassor car (1890) was of the midengine type (cf. Toyota MR 2/Fiera/Bertone 19), with the V-2 mounted transversely (à la Morris Mini/Honda/you name it). Panhard regressed the following year and placed the car's engine in the front, driving the rear axle—the formula followed by most of the automotive world for the next eighty years.

There existed in France exactly the right climate to launch a new industry. Technologists were cherished, there was plenty of investment money, and it was fashionable for the wealthy not only to have expensive toys but to be proficient in their use. A profound chauvinist joy wildly applauded the varied efforts of such pioneer aviators as Captain Ferber, Clément Ader, Gabriel Voisin, and others, and even embraced the Brazilian Alberto Santos-Dumont in his Parisian dirigible excursions. The same national enthusiasm was mustered to encourage the automotive exploits of individuals such as the Comte Albert De Dion, Louis Renault, and Armand Peugeot, as well as Panhard-Levassor and other established firms. There was all of this, plus the excellent tree-lined highways—Napoleon's autobahns.

The rest of the world was as susceptible as France to the intoxication of the automobile. In Great Britain, which would nurture some of the world's most luxurious and most eccentric automobiles, the way was paved by two Fredericks, Bremer and Lanchester. In Italy Aristide Faccioli designed the first Fiat in 1899, while the fine Belgian marque, Minerva, was launched in 1900. The magical name of Ferdinand Porsche first appeared in automotive literature in 1898 in Austria, where he did work on electric cars for Lohner. By 1905 he had moved to Austro-Daimler. In his future were speed records, Volkswagen Bugs, and Tiger tanks. In the Netherlands, the Spyker firm began building fine cars in 1900 and would continue to do so until 1925. In Russia even Tsar Nicholas became car-crazy; besides importing full-size luxury vehicles, he obtained a minicar for the sad hemophiliac tsarevich Alexis.

The Pieces Fall into Place

Suddenly, the world was ready, panting for liberation by these huffing, puffing, tarted-up carriages with their intoxicating sounds and smells. The acrid scent of horses and the clop of their hooves were soon replaced by the pungent fumes and the steady growl of ever larger, ever more powerful automobiles. A hundred smaller trades collapsed with the obsolescence of the horse, from village blacksmiths to small manufacturers of carriages, buggy whips, and surrey fringes. In their place grew thousands of gigantic industries: gasoline refineries; quarries for extracting the stone to use on endless highways; firms to make headlights, spark plugs, paint, tires, and all the associated paraphernalia of the vehicle.

This was only the starting point. Along the new highways subdivisions sprang up, as the automotive economic engine drove wages higher and higher until every worker could expect to own a car and a home. The character of cities changed, as commuting fostered urban sprawl; new species of restaurants and theaters—the drive-ins—added their bit to the social fabric and to the birth rate. The automobile changed virtually every aspect of domestic life, from employment to education, from vacations to vaccinations. And it changed the national character as well, for a nation brought up on automobiles is a nation prepared to drive tanks and fly airplanes in wartime. The world didn't know what had happened, and probably still won't for another hundred years.

A strange and wonderful romance between man and machine characterized the entire process. Of the perhaps 10,000 makes of cars that have emerged over the years, only a very few (for relatively brief periods of time) established a logical relationship between the performance people needed and the performance people sought. These sensible cars—some of them manufactured in the millions and loved by their owners —were apart from the automotive mainstream.

The public has instead—to whatever degree it could afford—opted always for speed, acceleration, road-holding, style, and luxury that far exceed any rational economic or emotional level. And it is this very lust for rolling life, the continuing search for the automotive pinnacle, no matter how unnecessary, that has brought most of the fun and most of the problems to the industry. In our pursuit of the rolling edge, we will follow these excesses with a fond but evaluative eye, and periodically we will speculate as to what in any given period could have been the ideal sensible car for everyone, contrasting it with the extravaganzas that were offered. What better place to start our journey than at the heart of the automobile, the engine?

The name Fiat has not always been associated with tiny, economy cars. This is Felice Nazzaro, in his Fiat Taunus, thundering on to win the Imperatore di Germania race in 1907.

C H A P T E R 2

THE POWER BENEATH THE ENDLESS HOOD

Too Much Is Not Enough

The modern automobile engine has both spoiled and intimidated us—and no wonder. We're spoiled because it so rarely breaks down. Given modern buying habits, many families never experience an engine failure in a lifetime of owning and driving cars. And we're intimidated because they've become so complex. Encumbered with emissions-control devices that could have come from a science fiction movie laboratory, they bully us with computers whose intelligence is contained in small chips and expressed in bizarre digital readouts or annoying voices.

If you haven't done so recently, lift up the hood of any current automobile, and try to name the various engine appurtenances you see. Certainly the fan is still a fan, but now it keeps running after the engine stops! And where on earth is the carburetor? The engine itself is turned 90°, so the spark plugs (if you can see them) run east–west rather than north–south. Various knobs, plates, boxes, and discs are interconnected with life-support tubes.

It will be difficult, but if you can find an antique —a Model A, say, or a stovebolt Chevrolet from the 1930s—lift its hood and look at the difference. The radiator is in front, unencumbered by a grille; the engine runs in line with the main axis of the car; the fan is easily visible; and you can find the starter, the generator, and the carburetor immediately.

Even the innards of the early cars made obvious sense; by looking inside the distributor, you could see that the lobe on the shaft rotated to separate the points, obviously creating the gap for a spark. The fanbelts were workmanlike, running from a pulley on the crankshaft directly up to the water pump—no nonsense here of having a fan run whenever it decides to.

We have brought the magnificent modern monstrosities upon ourselves, by responding to evolving problems with increasingly elaborate add-on solutions. Thus, each time we overcome an obstacle, we take another costly step into further complexity.

As a basic example, let's consider the power of the automobile engine. The early practitioners sought a replacement for the horse, and consequently the power levels were in single-digit multiples of horsepower. But as they gained proficiency, they succumbed to the overwhelming temptation to increase power—to 20, then 40, then 100, then (in the magic Duesenberg) 265! Think of it: 265 thundering horses under one endless hood, requiring not expensive bales of hay but inexpensive gasoline.

OPPOSITE:
A great photographer and a great car. The photo: Alfred Stieglitz's "Hand and Wheel." The car: a 1933 Ford V-8.

OVERLEAF:
Excursion in a Peugeot: In 1910, no one promised drivers a rose garden. ***(Photograph by J. H. Lartigue)***

Two views of the international favorite, the all-time classic 1936 Duesenberg SJ. Fitted with a gorgeous Rollston body, this convertible could reach 100 miles per hour in a sizzling seventeen seconds. When it came to detail, no point was missed and just as much attention was paid to the rear as to the front.

Why not keep adding power forever?

The reason why not came in 1973, with the largely bogus first fuel shortage, by which time we'd come to expect automobiles to be powerful enough to take us anywhere faster than we needed to go while surrounding us simultaneously with stereophonic sound, foam-rubber cushions, air conditioning, and dozens of other standard luxuries. We couldn't give these up; so to get the same (unnecessary) speed from more fuel-efficient engines, we began to raise compression, flirt with supercharging again, and carve off some of the purely stylistic unnecessary weight.

During this period, we also experienced a rising consciousness of the harm automobile exhaust fumes were doing to the environment. Consequently, at the same time that we were forcing ourselves to squeeze more power from smaller, less thirsty engines—when our cars were crying out for better engine breathing and access to the free air—we choked them tight with emissions controls. The final irony will come in ten or twenty years, when we discover that the vile new sulfurous emissions cause problems far worse than the old vapors we engineered out of existence.

The Internal Flame

One basic element of the internal combustion engine is the piston-cylinder combination—a working unit that can be traced to the pre-Christian era, when it was used by metallurgists to pump air into furnaces to intensify their heat. Things were simpler then, permitting an important step on the long evolutionary road to the modern internal-combustion engine to be taken by a man whose inventions presaged both aircraft and automobile. Hero of Alexandria, a geometer and philosopher on mechanical and physical subjects, saw that the power of steam could be used to create land vehicles and flying machines. His theories, as demonstrated by his Aeolusphere of about 150 B.C., were the first steps in a line of development that continues to this day.

Only two "short wheel base" (125 inches, compared to a standard Duesy's 142½ or 153½ inches) Duesenbergs were built. The first was purchased by Gary Cooper, the second by Clark Gable. This is Coop's car, bodied by LaGrande.

Hero was followed by other towering figures such as Evangelista Torricelli, a colleague of Galileo. Torricelli, also a mathematician, discovered in 1643 the principle of the barometer—the Torricellian tube. Torricelli's definition of the weight of air was incorporated into the thinking of Christian Huygens, a Dutchman, in 1680. Huygens, more famous for his propounding of the wave theory of light, designed an engine that used the explosion of gunpowder to reduce pressure in a cylinder, thus permitting atmospheric weight to move a piston. In a way, it is regrettable that this line of development was abandoned: think of the advertising possibilities of using 16-inch battleship shells as a source of power in a sports car, or of introducing high-octane nitroglycerine as a low-emission fuel. The internal-combustion engine of modern times and the primitive Huygens engine are of the same fundamental type.

Denis Papin, a Frenchman, worked with Huygens and later with the English physicist, Robert Boyle. He developed improved air pumps that relied on steam instead of gunpowder, and then in 1679 he invented a "steam digester" that generated high pressures; both advances were essential antecedents of James Watt's later steam engine.

In 1933, the name Cadillac had a sterling ring to it; if you drove one, you were clearly a member of American "gentry." Yet even among Cadillac owners there was a hierarchy, and the Imperial Limousine, with a body by Fleetwood, was its apex.

The maturation of the steam engine laid the true groundwork for the internal-combustion engine as we know it today. As the steam engine developed—first as the stationary element of the burgeoning industrial revolution, and then as the roaring rail engine of Richard Trevithick—significant improvements were made in lubrication, sealing, and metals. These permitted a number of experimenters to realize the fruits of Sadi Carnot's insight; in his 1824 publication, *Reflections on the Motive Power of Heat,* Carnot outlined the thermodynamic aspects of internal-combustion engine theory and even went so far as to postulate the corresponding aspects of what would be called the diesel engine.

There followed a series of incremental advances by pioneers who each added their imprint to the shaping of the incomplete concept. In 1833 W. L. Wright used pressure from combustion as a power source, and in 1838 William Barnett conceived the idea of compressing the charge of gases before exploding it. Alfred Drake introduced the "hot plug" (the forerunner of the spark plug) in 1843.

In 1860 Jean-Joseph Étienne Lenoir patented a stationary internal-combustion engine fired by the same gas that was used to illuminate houses. The engine essentially resembled contemporaneous steam engines, except that it was single-acting. In 1863 he installed an engine in a three-wheel carriage, and undertook an 11-km/7-mile drive.

Lenoir's practical efforts were followed by the theoretical achievements of Alphonse Beau de Rochas, who in 1862 formulated the now familiar "intake, compression, power, exhaust" cycle for internal-combustion engines. The concept of drawing a fuel charge into the top of the cylinder, compressing it with the stroke of the piston to form a highly explosive mixture, igniting the mixture with a spark, and then using the energy from the explosion to drive the piston—in order both to transmit power and to impart energy to a flywheel that would cycle the piston up again to exhaust the exploded gases—was revolutionary.

The next forward step was made by Alessandro Giuseppe Antonio Anastasio Volta—from whom we derive the term *volt*—whose experiments with electricity in 1777 led to the invention of a pistol that fired an explosive mixture of air and marsh gas (methane) by means of a spark. Next came efforts by Philippe Lebon in France, Isaac de Rivaz in Switzerland (who built and drove a vehicle powered by his internal-combustion engine), Luigi De Cristoforis and Guglielmo Marconi in Italy, and many others, all contributing to the scientific pyramid of collective reasoning that would reach its pinnacle in the work of Otto, Daimler, and Benz.

The engines developed by these last three men might erroneously be termed *simple*. Although less complex than modern types, they were far from simple, embodying as they did hundreds of years of innovative thinking, together with the earnest trial-and-error experimentation of brilliant inventors.

The Otto engine drew upon the principles of Beau de Rochas, using the intake/compression/power/exhaust four-stroke cycle. It was both heavy and low-powered, operating at about 120 rpm, but it nonetheless represented an important advance over steam competitors and was eagerly adopted by industry to drive machinery. The two-stroke engine, in which a scavenging and recharging process was substituted for the inlet and exhaust strokes, began being developed in 1878. It generated twice as many power strokes per cycle as did the four-stroke, but at a loss in effectiveness on intake and compression.

As noted earlier, Daimler improved on the Otto engine by increasing its speed. An automatic poppet, which served as the inlet valve, was arranged so that the vacuum created by the descending piston would suck it open. A camshaft operated the exhaust valve. The little 1.1-horsepower engine used a surface carburetor—one in which air was sucked across the top of a vessel of gasoline and drawn into the cylinders. Ignition was accomplished by means of a platinum hot tube used in place of today's spark plug.

The 462-cc-displacement engine was mounted vertically in a conventional carriage chassis; it had a single cylinder with a bore of 70 mm and a 120-mm stroke, thus standing as one of the first of the long-stroke engines that would dominate the industry for the next eighty years.

Karl Benz's engine was in some ways conceptually more modern than the Daimler. It was larger but generated only 0.75 horsepower at its slower 400 rpm. Mounted horizontally in a specially built three-wheel tubular frame, it had almost twice the displacement of the Daimler, at 954 cc. It used a similar surface carburetor to provide fuel through a side-valve inlet. This and the battery-operated trembler-coil ignition made the Benz engine more sophisticated than its Stuttgart competitor. The three-wheel Benz might also be considered somewhat superior to the Daimler because its chassis was tailored specifically to the engine, rather than being an adapted carriage. Yet the Benz engine was also concentrated on the basics: there was no radiator, and the cooling water was allowed simply to boil away.

Competition and Progress

In the first decade of the automobile, the number of cars and manufacturers grew slowly. The process was driven by amateur enthusiasts who already owned analogous businesses. Thus the manufacturers of wood-working machinery (Panhard-Levassor), bicycles (Peugeot, Duryea), and wagons (Studebaker) drifted into the new industry. Even so, by 1895 there were almost 500 cars in England and continental Europe, the preponderance of them either built by Daimler and Benz or derived from their ideas. In the United States, a

roughly equal number of automobiles probably existed, though from a much larger number of sources, for building an automobile seemed to suit every American blacksmith, machinist, and wagoner. There also seemed to be about fifteen promoters of cars for every one who made cars. And in the United States, the media took an important part in introducing both the automobile and a new sport, road racing.

Built in Putreaux, France, for George Kendall of Washington, D.C., this tire-less 1895 De Dion-Bouton certainly needed the services of the vulcanizer shop behind it.

Five years before the dawn of the twentieth century, the *Chicago Times-Herald* sponsored a 54-mile race from Chicago to Evanston and back that attracted ninety-eight entrants, most of whom were unable to field cars by race time. The entries

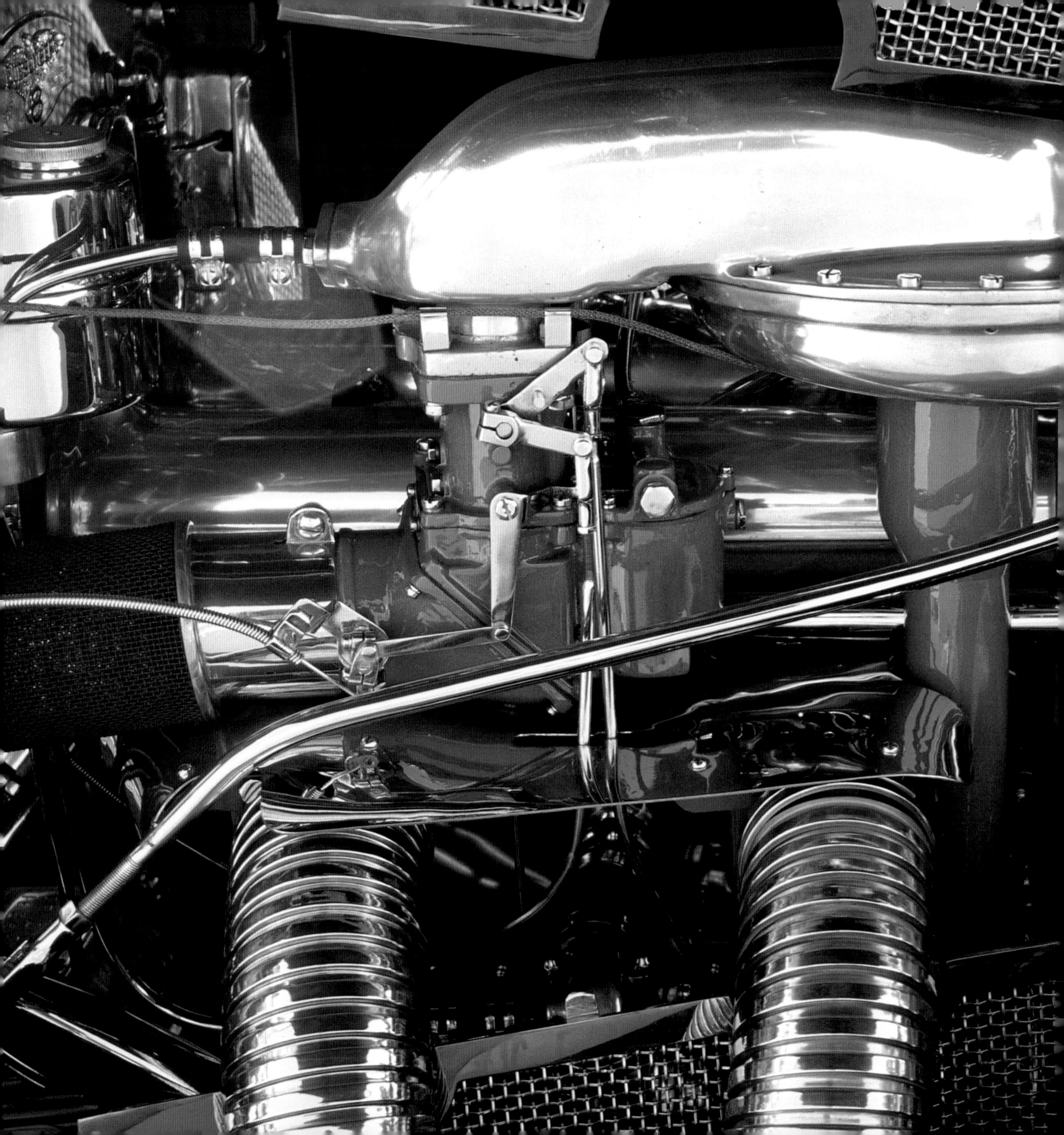

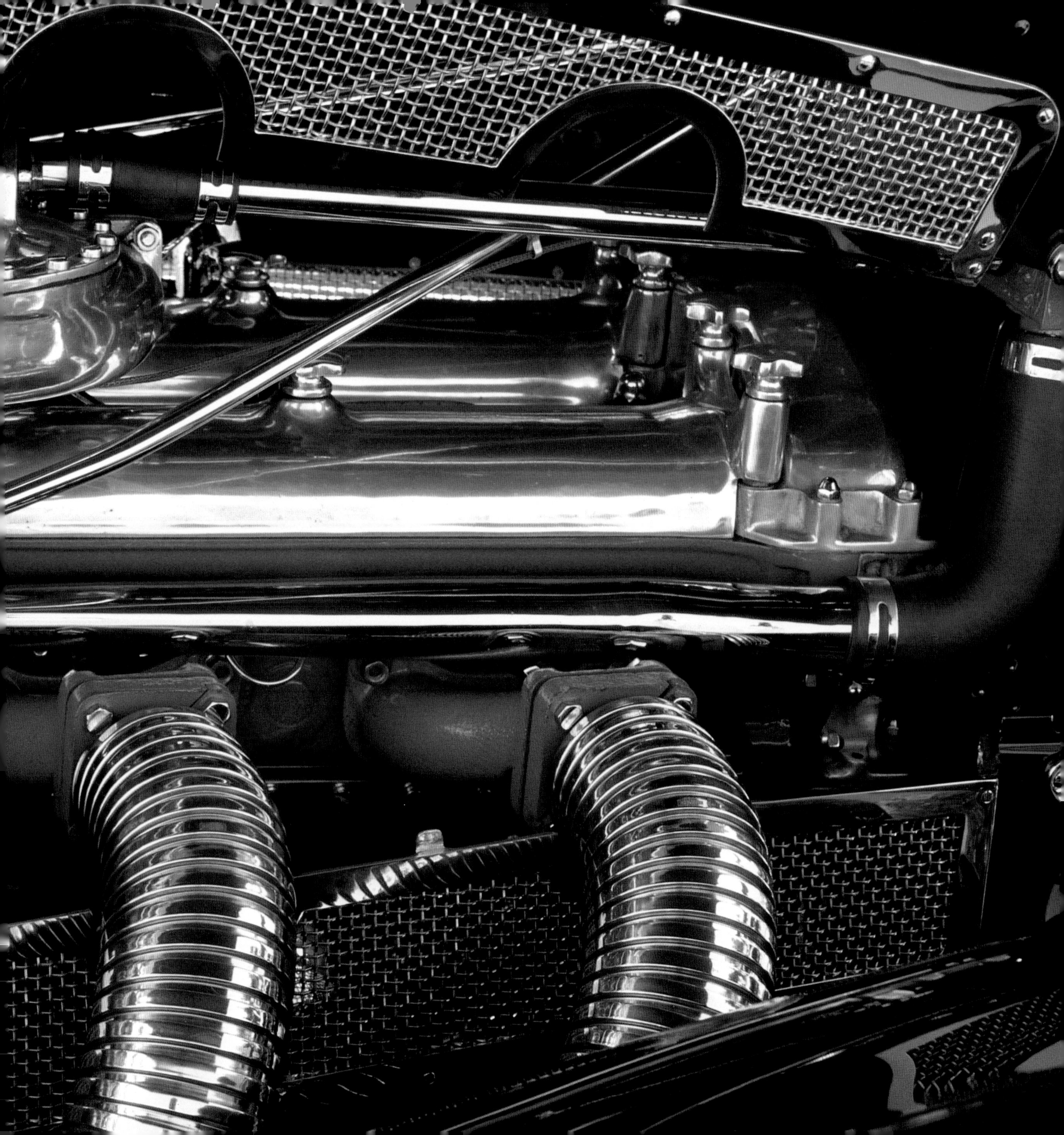

included gasoline, steam, and electric models, initiating a twenty-five-year larger contest to determine which type of engine would predominate in America.

The race rules were surprisingly comprehensive and seemed oriented more toward selecting a safe, practical vehicle than toward encouraging development of a racer. The real competitor of the automobiles in the race was the horse, for which the race officials were farsightedly seeking a mechanical substitute.

After a predictable number of difficulties, an arduous and well-managed race got under way on November 28, 1895, on a typical snowy Chicago winter's day. A Duryea won at an average speed of 7.5 mph—really remarkable, given the harrowing road conditions, and clearly better than a horse could have done.

The ultimate value of the *Chicago Times-Herald* race, however, was felt by the industry as a whole. Within three years, more than 200 firms were dedicated to the manufacture of automobiles; by 1900, national production reached 2,500 cars, a figure that would be matched two years later by the Oldsmobile company alone. Truly, an industry had been born.

Hiss, Bang, or Silence?

One automotive question that lacked an aeronautical equivalent was which type of engine to use—steam, gasoline, or electric. Early designers had advocated both steam and electric engines in dirigibles, but the power-to-weight ratio was simply impractical in aviation.

Steam had been successfully used for motive power in ships and railroad trains for years, and it was natural for it to be installed by tinkerers in early carriages. The firm of De Dion-Bouton offered steam automobiles—three-wheelers—for sale in 1888. The name *De Dion* remains current, as describing the type of independent rear suspension patented by the firm in 1893. Another early steam pioneer in Europe was Leon Serpollet, the builder of several hundred cars, one of which set a land-speed record of 75.06 mph in 1902.

The name that most people associate with steam cars, however, is *Stanley*—and for good reason. The bearded, twin Stanley brothers, Francis and Freeland, were unusual car-builders in that they came to the industry, not from heavy manufacture, but from a photographic business. They built their first steam car in Newton, Massachusetts, in 1897; by the spring of 1898, they had completed two—one of which they sold for $600. A series of demonstrations, culminating in a ramp-climbing contest, resulted in their receiving orders for more than 200 cars. Photography was promptly abandoned, and a legendary marque was born.

After a series of unsatisfactory but profitable deals involving the sale of their design, the twins reentered the business with a new design in 1902. In 1906, a Stanley Steamer (the streamlined Wogglebug) set an unofficial world's land-speed record of 127.66 mph. The firm prospered for the next decade; by 1917, however, improvements in the convenience of gasoline-powered cars—including the self-starter—drew unflattering attention to the inconveniences of steam.

Steam cars were powerful, vibrationless, and immensely quiet. They were also the devil to start and maintain, and operators needed a steam engineer's license as well as a driver's license. Start-up in the earlier versions could take as long as 45 minutes, and a steam car's water had to be replenished much more often than fuel in a gasoline car, typically every 40 to 80 km/25 to 50 miles.

Yet if enough genius and money were lavished on them, steam cars could be efficient. This was demonstrated beautifully by the lovely automobiles built by Abner Doble. Doble formed three companies in the decade from 1914 to 1923, each of which produced the best steam car in the world at the time. With astonishing success, he analyzed and solved the operational problems with other steam cars; a Doble could get up steam in as little as 90 seconds, and later models had an extraordinary range of 2,414 km/1,500 miles on one fill-up of water—fifty times the average of older steamers. But the car's engineering refinement was expen-

PAGES 50–51: **The engine compartment of the Duesenberg SJ, glittering with chrome and as well finished as the car's exterior.**

sive; with its luxurious Murphy body, it could cost as much as $11,200, in a day when an equivalent Cadillac went for $4,000. The fast, silent, luxurious Dobles attracted the interest of foreign and Hollywood royalty, but various troubles (including the Depression) beset the company, and only about forty-five Dobles in all were built before the firm quietly went out of business in 1931.

Later the great pilot/inventor/industrialist Bill Besler carried on the Doble tradition in cars and aircraft. Besler was a fascinating character who believed in ghosts, UFOs, and the inimical effects of fluoridating water, even as he made millions of dollars in agricultural equipment. He was the first (and, to date, the last) person to fly a steam-engine-powered airplane, in 1933. Besler later dabbled with the idea of reviving the Doble car in modern form, but the task was just too formidable.

Howard Hughes was also fascinated with the Doble, and set a team of engineers to improve on it to the typically extravagant Hughesian degree. To meet his performance requirements, his engineers had to run steam lines through the doors, where live steam could be counted on to scald the occupants in the event of a collision. Hughes reportedly examined it once, fired the engineers, and demanded that the car be destroyed.

Like the steam engine, electric motors had had a long and distinguished history, and they offered similarly quiet, vibrationless driving. They lacked power, however, and their thirst for recharging exceeded even the steam engines' thirst for water. Both steam and electric cars had troubles with long hills, which drained them of power and charge duration, respectively.

Electric cars were as easy to operate as steam cars were demanding. They were very reliable within the limits of their charge duration; a typical electric car was good for about 80 km/50 miles between charges. Perhaps most significantly, they were ''ladylike'' cars that seemed appropriate for women of the period to drive. For many, their greatest attraction was that they freed their operators from having to deal with a balky crank in those days before most gasoline cars had self-starters.

The manufacturers tried earnestly to escape the stigma of building only ''women's cars,'' since driving remained primarily a masculine activity. The electrics were the first closed cars, but this great styling advantage turned out to be rather chauvinistically discounted by the male buyer. Instead of appreciating the comfort, many prospective buyers considered the handsome closed bodies with large windows to be suitable only for people of the most delicate constitution and sensibilities.

The considerable popularity of electric cars declined rapidly after the introduction of the self-starter, and by the early 1920s they accounted for an insignificant percentage of the total number of cars on the road. The largest of the firms to emerge, Detroit Electric, stayed in business the longest, manufacturing electric cars hidden under the exterior metalwork of contemporary Dodges until almost 1940.

But the day of the steamer and the electric car may not be gone forever. With the specter of another fuel shortage looming, and with improved technologies in both disciplines, the great-grandchildren of the Dobles and Detroits may someday dominate the road. Gasoline and kerosene are convenient to use in a steamer, but conversions could be made to methane or (if worse comes to worst) wood- and coal-burning types. Similarly, improvements in batteries and in solar cells may give the electric a new career. Both types still have fervent fans, not unlike the hardy proponents of the dirigible, and when the time comes they'll be ready.

The Mainstream Emerges

By 1914 the internal-combustion engine had undergone a series of improvements that ensured its dominance in England and Europe; the same effect was evident in the United States within a few years thereafter. The reasons are clear: the gasoline engine was simple to build, easy to improve upon, durable, and capable of developing more

OVERLEAF:
The Bugatti engine compartment was as fastidiously finished as that of the Duesenberg, even though the actual engineering of the components ranged from thoroughly brilliant to maddeningly eccentric. This is the supercharged, eight-cylinder engine of a Bugatti 57SC Atlantic coupe. Filters were not commonplace, and the glass vessel is a sediment bowl for the fuel pump.

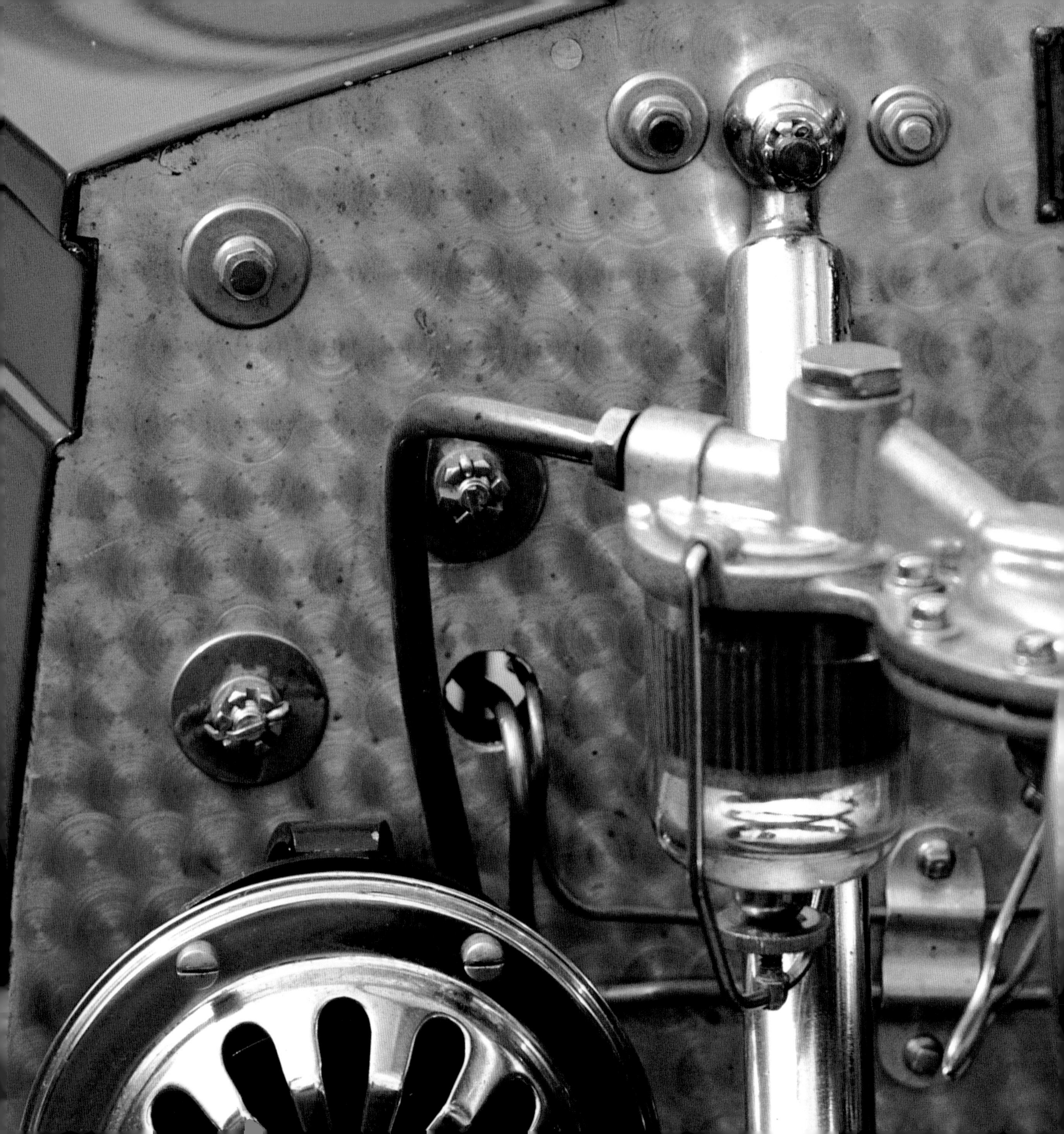

BUGATTI
BAS-RHIN
MOTEUR
POIDS
19 C.V.

than adequate power to satisfy the latest wave of new driving demands.

The early attraction of steam and electricity had been in part the result of their flexibility and smoothness. The gasoline engine, often a single hammering cylinder, was rougher and made stricter demands on the transmission of power; but the basic roughness of a single cylinder could be overcome merely by adding cylinders. The internal-combustion engine was susceptible to growth by various means, from total redesign (involving more and bigger cylinders) to simply bolting two smaller engines together. The bolt-on approach was later modified to a saw-off technique, when it became desirable to reduce engine sizes, but both methods underscore the docile responsiveness of the basic piston-engine layout.

A scene we may see again: charging the battery of an electric car. The Detroit Electric was manufactured from 1907 to about 1939, and pioneered the closed-car concept. Average touring range was about 80 miles between charges, but one record run of 211 miles was made.

In the very earliest days, engine placement varied widely. Builders could tuck the engine in front, in the middle, or behind, in keeping with their particular concept of the carriage it was to power. The automobile as an entity with a characteristic form of its own emerged with the Panhard-Levassor of 1891, whose Daimler-inspired engine was mounted forward and drove the rear wheels through a drive train that included a geared transmission offering three forward speeds.

Still-familiar names began to assert themselves in the widespread race to develop superior engines for superior motorcars. The awe-inspiring Mercedes marque came into existence in 1900 as a sales ploy, perhaps one of the most successful in history. Emil Jellinek, an aristocratic Austro-Hungarian diplomat, was an inveterate auto enthu-

The Stanley Steamer "Wogglebug" in which driver Fred Marriot (with goggles) ran a record 127.66 mph over the measured mile at Daytona in 1906. Francis E. Stanley (with beard) and brother Freeland O. were the guiding lights behind another concept that might yet return—the steam-powered automobile.

siast who used his contacts to sell cars to preferred customers. He had pushed the Daimler, but it fell into disfavor because of a series of accidents in local races—accidents attributed to the imbalance between the car's gutty power and its inferior road-holding capability.

Jellinek had worked personally with Daimler and Maybach to develop an advanced car with a huge 5,900-cc engine of 35 horsepower. The brute was installed in a chassis of equal sophistication (for 1900), and Jellinek knew he had a winner. To avoid the contemporary disenchantment with the name *Daimler,* he called the new car *Mercedes* after his daughter. It caught on. The modern Mercedes is still manufactured by Daimler-Benz AG. (The name *Daimler* was retained for the firm's products in England and Austria.)

Variations on a Theme

The Mercedes automobile was advanced in almost every respect, with a pressed-steel frame and a honeycomb radiator core. For the engine, the designers began the pattern of having the T-head cylinders cast as a unit and of calling for an integral crankcase. Both inlet and exhaust were mechanically driven, and an advanced Maybach carburetor was fitted. The engine was extraordinarily quiet for the time—so silent at its idling 500 rpm that people couldn't believe it was running. Its performance was spirited because it had a weight-to-power ratio of 15.2:1. In comparison, the very lively 1987 Mercedes 560S has to lug 21.9 pounds per horsepower.

The Mercedes had an enormous impact, caus-

TO

PAGES 58–59:
A Model T Ford competing in the 1909 transcontinental race from New York to Seattle. It won against fourteen competitors, in twenty-two days and fifty-five minutes, only to be disqualified later because it had stopped for an engine change en route.

LEFT:
Felice Nazzaro at speed in the Grand Prix of the Auto Club of France, June 26, 1912. When Napoleon began the wonderful network of straight, poplar-lined roads in France, he couldn't have foreseen that he was also laying the foundation for the sport of road racing. *(Photograph by J. H. Lartigue)*

RIGHT:
The 1922 Hudson Super Six used its full 76 horsepower to set a number of sales-engendering records in 1922, including a top speed of 102½ mph, covering a distance of 1,819 miles in 24 hours, and beating the rest of the pack to the top of Pike's Peak in 20 minutes and 5 seconds.

ing engineers everywhere to seek more and smoother power. Mere mobility was insufficient; cars now had to be smooth, tractable, and ever more innovative.

As advanced as the Mercedes four-cylinder engine had been, the type had a basic limitation: if displacement (a quantity computed by multiplying the diameter of the cylinder's bore times the length of the stroke of the piston times the number of cylinders) was increased too much, the four-cylinder engine became very rough and noisy. The first response to this limitation was to build more cylinders.

There soon appeared (in 1902) a special six-cylinder Dutch Spyker. It was the English Napier, however, that boasted the first production-line six-cylinder engine. The six was intrinsically smoother because the power impulse of the four-stroke cycle was no longer sequential but overlapping. Offsetting this advantage was the expense of creating a bigger block, additional parts, and most notably a longer crankshaft, which then proved to be subject to twisting. The answer to this problem was stronger crankshafts, flywheellike dampeners, and more bearings, but these additions escalated the cost of production even more.

The march to more cylinders was on. The sixes were followed by a straight-eight from C.V.G. and (in 1903) a V-8 by Clément Ader, the Frenchman who pioneered both the telephone and the airplane in that country. (Some French scholars still claim that Ader was the first to fly.) None of these cars was produced in quantity.

The first production straight-eight was manufactured by Dufaux of Switzerland in 1903—a great

12.7-litre engine that generated 150 horsepower. The cylinders were cast in pairs and surrounded by square water-cooling enclosures. Frederick Dufaux drove one to a world speed record of 97.257 mph in 1905. There were no airplane speed records at the time, but the Wrights had probably only exceeded the 40-mph mark. De Dion-Bouton built a small number of V-8 engines in 1907, but it remained for Cadillac—on its way to becoming the "standard of the world"—to produce the V-8 in quantity, beginning with the 1915 model. (Two banks of cylinders were inclined, so that from the front the engine had a "V" shape.)

Oddly enough, V-2, V-4, V-6, and V-12 engines preceded this epoch-making Caddie. Daimler had created the first V-2 in 1887; the French Mors made the first V-4 in 1898; Marmon displayed (but did not produce) a V-8 in New York in 1906, and Marion made a few V-12s in 1909. The currently popular V-6 first appeared in a 1911 Delahaye, while Packard reached a temporary limit with a twelve-cylinder Twin Six in 1912.

Motoring is not all cylinders and displacement, notwithstanding such excesses as the 17,657-cc Locomobile of 1906 and the 18,146-cc Fiat of 1908. No matter how fast these monster engines turned, each of the four cylinders seemed to be heard individually. (Fiat, in fact, followed the same Brobdingnagian approach with aircraft engines for a while, building huge powerplants of enormous displacement, before returning eventually to a more conventional scale.) With the main configurations well established, and with common-sense upper limits imposed on cubic inches and number of cylinders, it remained for engineers to refine elements of the engine for better performance.

Triumphant Benchmarks

Among the thousands of varieties of automobiles produced, hundreds have displayed brilliant engine design, but only a relative handful have so combined technology, utility, and spirit as to capture the imagination. In the survey that follows, the emphasis is on innovation, impact on the industry, and performance.

Cars and stars have always gone together. Here we catch an intimate moment as Joan Bennett and Gene Markey leave for the Hall of Records in Los Angeles to file their intention of marriage. The date is March 11, 1932; the car is unmistakably a Mercedes-Benz Model S.

THE ROLLS-ROYCE SILVER GHOST: OVERBUILDING TO PERFECTION

It is both logical and fitting that the next expression of automotive refinement would be the 1906 Rolls-Royce Silver Ghost. This inspired automobile —the prototype of which survives in pristine condition after more than 804,650 km/500,000 miles on the road—continued in production for twenty years. While the car was fitted each year with suitably current custom bodies of that period, its engineering remained basically unchanged.

The Rolls, like the Mercedes, benefited from a unique combination of personalities who set impossibly high standards and then achieved them. And like Daimler-Benz, Rolls-Royce had an early and sustained participation in aviation.

The honorable Charles Stewart Rolls, son of Lord Llangattock, was that most unusual of English amalgams: an aristocrat/sportsman who was also a fine businessman. A founding member of both the Royal Automobile Club and the Royal Aero Club, Rolls profoundly influenced aviation with the wonderful Rolls-Royce engines that powered many aircraft of World War I, and the Spitfires, Hurricanes, Lancasters, and Mosquitos of World War II. Rolls-Royce engines also won the Schneider Trophy, powered planes before and after the two world wars, and were equally dominant in civil aviation.

Rolls's influence was ultimately less engineering-oriented than personal. He was an ardent balloonist and aviator: the first Englishman to fly across the Channel, and the first man in the world to make the round trip. He donated to the nascent Royal Flying Corps its first airplane, and he sponsored men who would become great in the English aviation industry—A. V. Roe, Oswald Short, Geoffrey de Havilland, and many others. Ultimately he gave his own life in competition, crashing to his death in a Short-modified Wright biplane during a flying meet at Bournemouth in 1910.

His partner was a conservative genius named Frederick Henry Royce. Royce learned the value of rugged construction and high standards of quality as an apprentice on the Great Northern Railway, and he later established himself as an electrical engineer of the first water in his own firm, F. H. Royce and Company, Ltd. There he built electrical machinery, including dynamos and huge cranes. He was forty when his disillusionment with a French car, a 1903 Decauville, led him to design one to his own perfectionist standards. By 1904 he had built a far superior two-cylinder, 10-horsepower vehicle that, by word of mouth, attracted the attention of Charles S. Rolls.

The almost ideal combination of Rolls and Royce was bolstered by Rolls's associate, the estimable Claude Johnson, who had a brilliant flair for business. The similarity of this relationship to that of Daimler, Maybach, and Jellinek is noteworthy. The Rolls-Royce partnership, founded in 1904, was instantly successful in competition and in sales. Among the many excellent cars it built were some that departed from the norm, such as a successful three-cylinder model and the unusual Legalimit overkill, which had a smooth, powerful V-8 engine and a governed top speed of 20 mph.

The triumphant benchmark was the 40/50 model of 1906. The designation stemmed from a Rolls-Royce punctillio in defining horsepower. The *40* referred to the rating obtained by the Royal Auto Club formula, which was figured on the basis of the number of cylinders times the piston bore; the *50* was the actual brake horsepower of the engine. Later the 40/50's silence, style, and desirability elicited for it the haunting name *Silver Ghost* and set the pattern for future Rolls nomenclature.

The Rolls-Royce Silver Ghost engine set new standards of performance and reliability. The engine was composed of two cast-iron cylinder blocks of three cylinders each, cast with integral heads. This magnificent assembly was placed on an aluminum-alloy crankcase and further united by a seven-main-bearing crankshaft.

At 7 litres, it was a large engine, and it was square in that the diameter of its cylinder and the stroke of the piston were the same (4½ inches). In later years, the acceptance by the British government of the Royal Auto Club's formula for meas-

uring horsepower, for taxation purposes, would lead Rolls-Royce and all other British manufacturers to build long-stroke engines; and this would be one of the major determinants in differentiating American and British engine design philosophies. The Silver Ghost's compression ratio, 3½:1, was typical of the time, and a pump supplied pressure lubrication.

A side-valve system, known as the *L head,* was also introduced. Later it would be seen in millions of engines manufactured in every part of the world, but in 1906 the skill and standards of Rolls-Royce foundrymen were needed to cast engines with sufficiently accurate cooling passages to permit side-valve operation.

One of the all-time great misnomers was applied to this 1905 Dustless Spyker. The first with the six-cylinder engine and four-wheel drive, the Dutch Spyker obtained its greatest visibility as the runner-up to the race-winning Darracq in the charming 1952 film *Genevieve*.

The Silver Ghost was blessed with an impeccable jet carburetion system—something we take for granted today, but exceptional in 1906 when some cars still used wick or surface carburetors. In conventional cars, the driver had to coddle an engine through its acceleration and gear changes by means of constant manual adjustment of air and spark controls. The Silver Ghost made it easy.

Given Frederick Henry Royce's engineering background, it was natural that the 40/50 should have a magnificent electrical system. Two spark plugs were provided per cylinder, and these received their initial impetus from a coil; once the plugs were started, a magneto provided the spark.

A total of 7,876 Silver Ghosts were built in the

This is believed to be the first Rolls-Royce Silver Ghost to come to the United States. A Barker-bodied open touring car belonging to the Stevens family of Rome, New York, it is driven here by the chauffeur, James Stevens, ca. 1910.

course of their twenty-year production run (1,703 of these were produced in the U.S.). Changes were introduced over time, in a manner that became the Rolls-Royce theme. Rolls-Royce chose not to be in the vanguard of change; it simply made sure that its older methods were superior to any newer ones until it had time to develop the best of these to a comparable standard of excellence. Thus, four-wheel brakes were adapted from Hispano-Suiza practice in 1923, a longer-stroke, overhead-valve engine was introduced in 1925, and independent front-wheel suspension made its appearance in 1936. The process continued over the years.

The pleasure of driving the Silver Ghost was only part of the attraction in owning one. The attention to quality extended to every detail, including the use of lighter, high-strength alloy steels to reduce weight, which kept the car's fuel consumption to a remarkable 29 km/17.8 miles per Imperial gallon. Like the contemporaneous Mercedes, the Silver Ghost was a top competitor, whether laden with a limousine body or stripped down to perform as a two-seater.

Over time, competitors to the Rolls-Royce would emerge: Leyland in England, the Hispano-Suiza from France and Spain, Italy's Isotta-Fraschini, and the American Duesenberg. Each would have its adherents, but none would survive as the Rolls had survived, and none would bear the implicit assurance of perfection that is the Rolls tradition.

THE FORD MODEL T: QUALITY SUFFICIENT FOR THE MASSES

There are surprising similarities in the histories of two cars built for radically different purposes, the Rolls-Royce Silver Ghost and the Ford Model T. Henry Ford's background and apprenticeship was the American equivalent of F. H. Royce's. Their dedication to the path selected was the same: Royce sought to find the very best way to build the most durable and best-performing car for those who could afford it; Ford sought to make a good automobile affordable to the greatest number of people. In each process, many convergences of technique occurred. Both firms settled on making one product. Both tended to integrate production vertically, although Ford followed this to an obsessive degree, while Rolls-Royce moved away from it in later years. The closest similarity might be in their respective adoption of a process of progressive change. The Model T is sometimes thought by laymen to have been conceived as an entity in 1908 and manufactured without improvement through 1927, but nothing could be further from the truth. Engineering changes were introduced every year, in addition to periodic styling updates.

The two firms were both involved in aviation, although in totally different ways. Rolls-Royce built aviation engines continuously. Ford built Liberty engines during World War I and Consolidated B-24s and a variety of other products during World War II. In between, Henry Ford sponsored an airline and caused the famous Tri-motor to be manufactured. Today, divisions of both companies are engaged in the high-technology aspects of aerospace engineering.

The difference in marketing philosophy of the two car manufacturers meant that the engines must be very different. The Rolls-Royce engine was virtually custom-built, hand-balanced, and intended for a lifetime of use. The Model T engine was mass-manufactured; at the peak of production, an entire car took only a little over one hour to build. The Model T used high-quality materials by contemporaneous standards, but the design itself was the essence of simplicity.

With about one-third the displacement of the Silver Ghost at 2,896 cc, the Model T's L-head engine had a 3¾-inch bore and 4-inch stroke. The carburetor was made by Holley, and the ignition was a combination of coil-start and magneto for current. Later models had conventional coil, battery, and distributor arrangements. All of 22 horsepower were generated at 1,600 rpm, but the light weight of the car allowed lively performance.

Not for the Model T were such Rolls-Royce niceties as full-pressure lubrication or water pumps.

TOP:
Travel in the Model T provided many a spontaneous opportunity for a leisurely view of the countryside while repairs were being made. ***(Photograph by Arthur Rothstein)***
BOTTOM:
The Model T could also do a prodigious amount of work, including moving haystacks when required.

Lubrication was by splash and by god, and a thermo-siphon system was used for cooling. The Rolls-Royce crankshaft was supported by seven lovingly built main bearings, pressure-oiled and massively redundant. The Model T's engine had three babbitted main bearings of limited durability; their replacement was such common conversational fodder that the name *Babbitt* (so suggestive of malleability) was used for the eponymous hero of Sinclair Lewis's 1922 novel.

And where Charles Rolls and Henry Royce had looked to the heavens for inspiration, seeking to sell their motorcars to royalty and to select members of the gentry, Henry Ford sought his inspiration on the good green earth. He wanted to build cars by the hundreds of thousands; his product and his philosophy resulted in their being built by the millions. As production went up, prices came down; as prices came down, production went up. He threw another famous variable into the equation by raising wages to an unthinkable $5 per day, and by employing people Detroit had previously considered unemployable—minorities and the handicapped. His motives were not entirely altruistic: he wanted everyone to earn enough to buy one of his cars. In 1909, the Model T cost $850 for a touring car, and 17,771 were sold. In 1923, 1,817,891 were sold, at prices as low as $290.

Ford's little four-banger was easy to repair; it was also easy to soup up for increased performance. The engine was ideally suited for its time and came as close as any engine ever has to being the basis for a true people's car.

Ford was not alone in his search for an economical, serviceable car for the masses. The role of his Model T was filled in England by the Austin 7 and the Morris Cowley, and in France by the Peugeot Quadrillette and the Citroën 5CV; and similar tiny cars appeared in many countries—most of them built under license to Austin or Morris. But there was an essential difference in quantity. In the seventeen years of the Model T's production (from 1922 to 1938), only 375,000 Austin 7s were built—a spring production run for the Model T in its day.

THE CHEVROLET STOVEBOLT SIX: THE STATUS-SEEKER'S CHOICE

The Model T paved the way for a host of imitators, competitors, and follow-ons. The Tin Lizzy was the kindergarten of motordom, the entry level to auto-propped Ego School. A young man might start out his career driving a Model T, but as success came he wished to make it evident to others by driving a ''better'' car.

And there were many manufacturers eager to provide him with the opportunity, including William Crapo Durant, one of the most flamboyant of the automobile magnates before John DeLorean. Durant brought the Chevrolet into the world, putatively to be a head-on competitor to the Model T. His real motive, however, was to use the new car as a means to regain control of the giant corporation, General Motors, he had founded in 1908 and then lost control of in 1910. He succeeded in 1915, only to be bounced again in 1920, when a combination of the recession and his own profligate business ways did him in. It was Alfred Sloan, the later genius of GM's greatest successes, who turned the Chevrolet from its course of suicidal competition with the entrenched Model T to being just a cut above the Ford in terms of consumer desirability.

Chevrolet had gained first place in sales in 1927 and 1928, as Ford ceased production of the Model T and struggled through its convulsive conversion to the Model A, but Sloan played his trump card in 1929 with the introduction of the all-new Chevrolet International, powered by a 46-horsepower, six-cylinder engine.

The new six-cylinder engine was refined for the time, with overhead valves, an almost square engine of 194 cubic inches, and a power output of 46 horsepower at a time when the four-cylinder Model A engine was putting out 40. In truth, though, there was no vast difference in comparable models of the two firms in terms of price, weight, or performance. The six was noticeably smoother at lower rpm rates, but the important thing from a promotional standpoint was that the CHEVROLET was a *SIX*, while the Ford was a *four*, and everyone knew what that meant.

Over the next twenty-six years, the Chevrolet Six was maintained in production with only minor refinement—the compression ratio was raised over time, and in 1937 the number of main bearings was raised to four—and it mattered not a whit to sales. The car sold on the merits of being stylish, ''solid,'' and (demonstrably) not a Ford. It offered entry level to the General Motors ladder of symbolic affluence; and if Cadillac provided a new creature comfort or styling motif, it would ultimately filter down to the Chevrolet level. The engine remained essentially the same (that is, adequate) over the years.

The success of Sloan's Chevrolet philosophy was confirmed in sales results. Inertia and the loyalty of Model T owners who had eagerly anticipated the Model A regained Ford's customary number-one position in sales in 1929 and 1930, but after 1931 the sales leader was always Chevrolet. There were two exceptions: Ford eased Chevrolet out in 1935 by the sheer force of its V-8 models' beauty, and again when the war truncated production in 1942.

The Chevrolet Six was simple to maintain, economical to run, and rugged enough to last the expected lifetime of an American car—four years or 60,000 miles, whichever came first. There were exceptions to this term of useful life, but not many; Detroit's philosophy was, ''if you can't wear them out, rust them out.'' Those big companies didn't invest in planned obsolescence just so the motorist could drive an old car!

The Chevrolet Six was a significant step along the pell-mell path of planned obsolescence that manufacturers continue to follow to this day. If Durant had been a philanthropist and a philosopher rather than a supersalesman and industrialist-brigand, he might have brought forth an improvement on the Model T engine—one that would run for 200,000 miles—and he might have installed it in a car that did only what people needed, not what they wanted. And we probably would never have heard of him.

V-8s: ROYAL AND POPULAR

It is hard to imagine the excitement caused by the introduction of the six-cylinder engine in the 1929 Chevrolet: Overnight there was a car, comparably priced to the Ford, that offered the luxury of six cylinders in a body style obviously related to the more expensive cars in the General Motors line. This is a four-door phaeton, of which 8,632 were produced at a base price of $525 each. In this condition today, it is valued at approximately forty times that amount.

The automobile market was stratified. There was always a large clientele that could afford almost whatever it wished, and a somewhat smaller one that could in fact indulge any desire whatsoever. The latter class was the one most appealing to artist-engineers such as Ettore Bugatti, Ferdinand Porsche, Arthur de Connick, W.O. Bentley, and Fred and August Duesenberg. The liberty to design something as elegant as imaginable and as complex as necessary, without regard to purchase or maintenance costs, was intoxicating.

But the larger group of car buyers—those who could afford *almost* anything they wished—provided an attractive target for many manufacturers. The first law of motordom dictates that there is a larger profit margin on an expensive car than on a cheaper one. The second law is that, in a mild economic downturn, the rich will stay rich enough to buy what they want. Unfortunately, there is also a third law, and it is responsible for the demise of such great names as Marmon, Peerless, and Pierce-Arrow: in really bad downturns, even the rich stop buying long enough to destroy companies.

The element of profit in the history of car manufacture is quite remarkable. In a book written in happier times, *On a Clear Day You Can See Detroit,* John DeLorean makes the point that, while a Cadillac cost GM only a few hundred dollars more to build than a Chevrolet, it sold for several thousand dollars more.

It was to appeal to the larger class of moderately well-off people that Cadillac became the first marque to produce a V-8 in quantity—a decision that again showed automotive cross-

pollinization. Henry Martyn Leland, whose white hair and goatee gave him a patrician appearance consonant with his products, had been a key figure in the success of first Ransom Olds and then Henry Ford. Farm-born but machinist-bred, Leland cut his engineering teeth on the sharp standards of the Springfield Arsenal and Samuel Colt. From them he learned the value of precision and interchangeability of parts—two things that would make inestimable contributions to Cadillac, which he helped found in 1902.

In 1908, the Cadillac was demonstrated in London in devastating fashion by F. S. Bennett. Bennett had three Cadillac runabouts selected at random from the eight that were available to him, and shipped them to the famous Brooklands racetrack. There they were completely torn down, and their parts were thoroughly mixed. To introduce an additional element of spice, spare parts from the London depot were substituted at random. The cars were then reassembled—their mixed ancestry dramatically demonstrated by their Joseph's coat of many colors—and driven for 805 km/500 miles around the banked Brookland track. Speeding at 34 mph and averaging 30 mpg, the cars concluded a tour de force that gave Cadillac a basis for its 1913 motto, "The Standard of the World." It had earned the new motto for a number of reasons, not the least of which was Charles F. "Boss" Keterings's self-starter, introduced in 1912.

Cadillac had become a part of General Motors in July 1909, and it was inevitable that Leland and his son Wilfred would have a dust-up with William Durant. The split came in 1917, when Durant declined to undertake production of the "war-winning bully Liberty engine." The feisty senior Leland, 74 at the time, had no trouble starting the Lincoln Motor Company and garnering orders for 6,000 Liberty engines that would in part inspire the later Lincoln V-8.

The Cadillac V-8, which was introduced in 1915 at the bargain price of $1,975, had a similar aviation heritage. The design team, headed by Wilfred Leland, had evaluated a number of automotive V-8s, including the De Dion-Bouton, but

Ford V-8 cars were an important part of Bonnie Parker and Clyde Barrow's "business," to which Bonnie's smile *(opposite)* and Clyde's letter to Henry Ford *(right)* readily attest. *(Photograph of Bonnie Parker by the Barrow Gang)*

Tulsa Okla
10th April

Mr. Henry Ford
Detroit Mich.

Henry Ford
RECEIVED
APR 13 1934
Secretary's Office

Dear Sir:—
While I still have got breath in my lungs I will tell you what a dandy car you make. I have drove Fords exclusivly when I could get away with one. For sustained speed and freedom from trouble the Ford has got ever other car skinned and even if my business hasen't been strickly legal it don't hurt eny thing to tell you what a fine car you got in the V8 —
Yours truly
Clyde Champion Barrow

was particularly interested in the Hall-Scott airplane engine whose design had greatly influenced that of the Liberty. The engine was an L-head V-8 with a displacement of 314 cubic inches; it had a 3⅛-inch bore and a 5⅛-inch stroke. The block was cast in two four-cylinder halves, and joined at a 90° angle. A gear-type oil pump pressure-fed the oil, and a power tire pump was mounted within the V.

The 70-horsepower engine situated in only a slightly modified version of the preceding year's body provided a new basis for Cadillac advertisements, which spoke proudly of "The Penalty of Leadership." It would be refined over the years into a smooth, powerful V-8 that stood as the supreme symbol of American automotive luxury.

LUXURY COMPETITORS

The success of Cadillac's V-8 appealed to some competitors and appalled others. It appealed to Packard, Peerless, Stearns-Knight, Cunningham, and others that could quickly respond with their

own models (Packard's being a V-12). It appalled Locomobile, Marmon, McFarlan, and others that lacked the resources or the inclination to compete. Olds came out with a V-8 in 1916, and even Chevrolet introduced one in late 1917, only to discontinue it the following year.

Perhaps the greatest influence of the Cadillac V-8 came some years later, in the Depression-era race for prestigious customers. When bad times came, companies invariably tried to solve their problems with bigger, more powerful engines. Cadillac itself upped the ante with a V-16 in 1930 and a slightly more restrained V-12 in 1931. Marmon succumbed to this madness with a V-16; Peerless did the same; and Pierce-Arrow went out grimly holding onto a V-12.

These larger powerplants were often engineering masterpieces made of the best materials and capable of remarkable performance. A silver-and-black 1933 Pierce-Arrow convertible averaged more than 113 mph for twenty-four hours at Bonneville. Famous record-setter Ab Jenkins pushed it to average 125 mph for the last hour—during which time he lathered up and shaved himself with a safety razor, to demonstrate the smoothness of the ride at high speed. But multicylinder monsters weren't relevant to the existing economic situation—and probably weren't relevant to any other, either. More than forty years after the Cadillac's debut, however, equally unnecessary V-8s would be established as the rule and not the exception.

In the meantime, European prestige cars continued to be equipped with perfectly suitable straight-six and -eight engines, as did Chryslers

An elegant 1930 Cadillac 353 Hibbard et Darrin Berline Transformable, as luxuriously fitted inside as it is immaculate out.

OPPOSITE:
One of the most desirable of all classics, the exotic Bugatti 57SC Atlantic-Electron. The stylish coupe was originally designed to be built from magnesium, which required riveting body sections together. Aluminum was used instead, and though not required, the rivets were retained for aesthetic reasons.

BUGATTI

A 1927 Hispano-Suiza, the car of sporting kings. *(Photograph by J. H. Lartigue)*

and lesser makes. The real difference in performance of a V-8 over an equally well-made six or straight-eight ranged from negligible to nonexistent, but the buying public wouldn't believe it. And if V-8 power was good in a Cadillac, why wouldn't it be good in a Ford?

V-8s FOR THE MASSES: FORD'S ANSWER TO CHEVROLET

Henry Ford was a simple man. He simply wanted to sell more cars than anyone else, and he simply wanted to build them exactly as he chose. But times had changed since he built the first Model T. Ford had been annoyed by World War I, and the Depression was another affront. But being second to Chevrolet was worst of all. By 1930, he had committed his firm to the biggest gamble in automotive history, one that came within an ace of destroying Ford the company and Ford the man: he resolved to create a low-cost, mass-produced V-8 for 1932. It was as revolutionary as if Lee Iacocca had determined to put a gas turbine in the K cars. Two years was not much time to create

a conventional engine; to design and build a V-8 requiring extraordinarily complicated castings in order to be produced in quantity was next to impossible. But with his usual drive and his uncanny knack for picking the right people to do the work, Ford succeeded. He had at his disposal the world's largest automobile plant and one of the world's largest research facilities, yet he forced the small team to work in a replica of Tom Edison's Florida laboratory. While the setting was perhaps inspirational, it meant working without instruments, without electric motors for tools, and without even a test stand upon which to run the engine.

But how can you say he was wrong? The small team, headed by Emil Zoerlin, Carl Schultz, and Ray Laird, slaved under Ford's direct supervision, in complete secrecy from the rest of the company. Somehow—probably because of Ford's obsessive interest—they fulfilled his demands, and a $300,000,000 rush project ensued to get the engine into production.

The engine itself would be produced with only minor refinements for the next twenty years. It was a flat-head V-8, of monobloc (that is, one-casting)

construction, with a 221-cubic-inch displacement. It generated 65 horsepower—5 more than the competing Chevrolet.

The rushed program was perhaps the first sign that Ford, approaching 70, may have been losing his automotive grip. The V-8s debuted during the worst production year in American automotive history, and there were pinhole flaws in the casting that required massive recalls. In addition, to canny Americans, road tests demonstrating rather economical gas consumption rates of 20 miles per gallon for the V-8 didn't overcome the certain knowledge that eight cylinders would require more fuel than four or six. Ford's company stayed in second place behind Chevrolet, with sales of just over 250,000 cars. Yet the engine he wrought maintained itself in progressively improved form over the next twenty-two years, until the debut of the high-compression, overhead-valve Ford line in 1954. The original V-8 was changed only slightly in this time—the displacement going up to 239 cubic inches, and compression ratio rising from 5.5:1 to 7.2:1. These small tweaks, combined with increasing the rpm at which the rating was taken to 3,800, resulted in a competitive 110 horsepower being generated.

The luxurious 1930 Franklin Pursuit *(above)* was air-cooled, but its front end was still cloaked in a grille, like every water-cooled car of the period. A Hudson *(below)* from the fine 1950 series.

THE V-8 BECOMES STANDARD

The new era of V-8 engines was embodied in

the sensational Oldsmobile Rocket 88s of 1949. ''Futuramic'' styling had debuted in the top-of-the-line Olds 98 in 1948, but the new overhead-valve V-8 engines and their 135 horsepower set the automotive world's wheels spinning.

A number of things coincided to make these engines more noteworthy than might have been expected, given that Ford had been flogging V-8s in Fords, Mercurys, and Lincolns for years, as had Cadillac in its cars. First, the new Oldsmobile styling was exciting and made the cars look as fast as they proved to be. Second, all cars were getting heavier and more equipment-laden. Air conditioning, which had first appeared on a 1940 Packard, was just one of the elements draining power and adding weight to the new cars; stylists at all companies were dictating more room, more chrome, and more overhang. Third, racing, including road racing, had resurfaced after the war.

Like Hudson, Nash built great cars for years until public pressure for rapid model change was augmented by the demand for ever-higher-horsepower V-8 engines. Smaller companies could not compete with the Big Three, and one by one they disappeared. But when this Nash 482-R coupe was on the road, people turned to look. Today, they still do.

But fourth and perhaps most important, for the first time in America, automobile performance reports began to enjoy wide popular circulation. Accurate, if generally friendly reports, had long been made in English magazines, but Tom McCahill, writing colorfully in *Mechanix Illustrated* magazine, started the trend toward popularization of car stats and facts that continues unabated in the United States and Europe today. McCahill's analysis was usually kind and was expressed in a sort of code. He rarely said that an automobile's attributes were terrible; some cars were just better than others. And throughout he emphasized acceleration times, road-holding, quality of fit, and other performance criteria that had never been reported with any degree of objectivity before. Readers loved the personal touch—it made them feel knowledgeable, too.

A boomlet of magazines swelled to capitalize on this new American interest, and out of it were born *Motor Trend, Road and Track, Car and Driver,* and many others, which, along with *Consumer Reports,* have raised the appraisal of automobile performance—if not to a science—at least to a discipline.

It is a mixed blessing. Most of the magazines (*Consumer Reports* excepted) focus on degrees of performance that most ordinary drivers don't need, can't attain, and wouldn't recognize. Yet precisely these performance levels—''0 to 60 in 8 seconds, 0.8 G on the skidpad, top speed of 140 mph''—are the siren song that sells cars and dictates the engineering paths of giant motor companies.

On the other hand, the effect of these magazines has been to shake Detroit out of its condescending stupor by focusing attention on the superior handling, economy, and general performance of European and Japanese cars. If it has been carried too far by an excess of zeal, that is because the participants are enthusiasts and because counseling automotive temperance has never built a magazine's circulation.

And while it was the fuel shortage that eventually spelled the end of the big-displacement, huge-horsepower cars that Detroit had conned Americans into loving, these same magazines made the new generation of smaller and more efficient vehicles not only palatable but desirable. Government regulation may have led to downsizing and other requirements for fuel economy, but the automobile magazines proved in print that better cars would result.

The net effect of the sprightly performance of the V-8s was first to drive companies like Nash, Studebaker, Packard, Hudson, Kaiser-Frazer, and even Chrysler to the wall to develop comparable V-8s. Other market forces were also operating: styling became more important than ever, and the sheer volume of General Motors production made its cars look the standard, no matter how bloated and vulgarly bechromed they became. But efforts by Hudson to tart up its excellent large six with Twin-H power or by Studebaker to drop big Packard engines into light Stude chassis simply weren't enough. Hudson's styling, fresh when it first appeared in 1948, seemed to grow as bulbous as a retired wrestler by 1954. Kaiser-Frazer, which had introduced striking beauty into its 1951 line, turned to supercharging in a desperate effort to jazz up its solid, stolid Kaiser Continental six, but to no avail.

Oddly enough, many of the V-8s produced by the smaller companies were excellent. The Packard V-8, introduced in 1955 in a Richard Teague–styled body was splendid, but by then the company was already locked into its moribund merger with Studebaker. In 1951, while still on its own, Studebaker had built a very fine V-8. The newly formed American Motors Corporation used Packard engines in both members of the Nash-Hudson line of 1955, and then introduced its own sound V-8 in 1956. For most of these companies, the V-8 was a problem, not a solution. American Motors survived because George Romney and others saw that the solution lay not in more cylinders (no matter how aligned), but in smaller engines.

In a final bit of competitive irony, the V-8 engine that had the greatest impact on the minds of Americans—and especially of hot-rodders—was not a Cadillac, a Lincoln, or even a Ford. It was the overhead-valve Chevrolet engine, introduced in 1955 and the progenitor of a development line that continues to this day. The engine was a combined marketing and engineering ploy conceived at the very top of GM's management by Harlowe Curtice and entrusted to the design talents of Ed Cole and Harry Barr. These men created a lightweight, 265-cubic-inch oversquare engine that put out 162 horsepower at 4,000 rpm. The Chevy V-8 received instant and continuing acclaim at a level rare for a stock mass-produced engine, and it was

probably subjected to more intensive souping up by hot-rodders than any other engine in history.

This sophisticated dual-overhead-camshaft V-8 belongs to a 1985 Ferrari GTO; it has 32 valves and generates 394 horsepower at 7,000 rpm. Don't try to repair this with a screw-driver and a Chilton's manual.

OTHER ENGINES OF NOTE

It's ironic that a country guilty of ignoring its own remarkable air-cooled Franklin would respond with such fervor to the Volkswagen. The reason for this disparity, of course, was that the Franklin's air-cooled engine—excellent though it was—was installed in an otherwise conventional automobile. The VW, the blessed Bug, arrived at a time when a small car could be both economical and chic. It was clearly the right car at the right time.

Sadly, the VW's flat-four engine, which delivered both spirited performance and excellent fuel economy, was not amenable to emissions-control devices, and so had to be replaced.

The Bug was not alone. The smooth, durable, in-line six- and eight-cylinder engines that had been used so successfully in cars with many other familiar names—Plymouth, Dodge, De Soto, Chrysler, Pontiac, Buick, Nash, Reo, Hudson, and Hupmobile, among others—disappeared on the twin tidal waves of fashion and economy. Some of these were magnificent creations: the Nash sixes, for example, were smooth, silent, and durable; and the Hudsons were good enough to be selected by Reid Railton to be used in his line of expensive English custom cars.

Yet these were all mass-produced cars. The most interesting engines—and often those in the forefront of development—came from exotic cars with magical names: great roaring masterpieces that flouted tradition and flaunted bank accounts with equal abandon.

These wonderful engines were nurtured by a strange mixture of genius, arrogance, and disregard for economics; and often they bore signs of a deep association with aviation. Ettore Bugatti, who began his work in 1902, was the genius behind a colossal twin-eight aircraft engine that

A side view of the 1930 Franklin Pursuit. Its paint job and trim are unusual for the period.

was intended to generate 500 horsepower and permit a 37-mm cannon to be fired through the propeller hub. The engine was just too extreme for production in the rush days of war, but it served as a fertile source of inspiration both for later Bugatti engines and for those of Fred and August Duesenberg.

Like many automotive and aviation pioneers, the Duesenbergs' roots ran to a bicycle shop, and a bankrupt one at that. But a dedication to the highest standards led the Duesenbergs to build engines that won recognition on the pre–World War I racetracks, putting the two brothers in a position to work on the Bugatti monstrosity. They looked and learned, and their postwar engines were triumphs—half-scale, eight-cylinder, overhead-camshaft derivatives of the too-complex Bugatti. They placed these in an upscale automobile that featured, among other things, four-wheel hydraulic brakes. The brothers Duesenberg lacked business acumen, however, and it was not until they came under the benevolent wing of Errett Lobban Cord that the phenomenal J and SJ Duesenbergs materialized.

The J was introduced in 1928, featuring twin overhead camshafts and four valves per cylinder, and generating 265 dynamotor horsepower. In 1929, $8,500 bought only a chassis; customers could then either purchase a Gordon Buehrig in-house-designed body for $6,000 more, or pay as much as they wished by selecting from among the custom coachmakers of the world. As a comparison, a Cadillac Fleetwood with a 90-horsepower V-8 cost about $4,200. Just before his tragic death in one of his own automobiles, Fred Duesenberg designed the supercharger that would raise the horsepower ante to 320 in the SJ series.

Only about 500 J or SJ Duesenbergs were built, but they had a disproportionate effect on automotive history. They were clearly the only Ameri-

can car that could claim the Rolls-Royce as a peer; and no other car can claim to have generated a comparably familiar and long-lived phrase as ''It's a Duesy,'' to mean the very best.

Ultimate Engines

As in the Duesenbergs, a racing heritage is echoed in the world's other most exclusive and expensive marques. The very name *Aston Martin* derives from the famous win of Lionel Martin in the 1914 Aston Clinton hill climb in a 10-horsepower Singer. Today's Aston Martin's horsepower is discreetly unstated; if you purchase a £74,200 example, however, the factory will confide in you. In recent years, Aston Martins have grown in size somewhat and the firm has been purchased by Ford. Their place in the grand prix racer category has been assumed by the Lotus Esprit Turbo, a four-cylinder car capable of speeds approaching 150 mph.

Billed as the ''Car of the Future'' at the New York Auto Show, in 1941 the Newport Dual Cowl Phaeton was the first non-production car to serve as an Indianapolis Speedway Pace Car.

The legendary racing name of modern times is, of course, Enzo Ferrari, who both raced and built racecars before beginning to build the wildly successful line of Ferraris in 1946. Their enormous success in grand prix racing created a demand that drove production steadily upward to about 2,500 per year. Various engines and body styles were produced, but the name *Ferrari* always seems somehow to evoke great, sonorous twelve-cylinder grand prix cars. Even Ferrari had to bow to the exigencies of conservation, and the last V-12 appeared in 1976 in the 400i; but in 1984, a brilliant new design, the Testa Rossa, was brought out in response to popular demand. The name refers to the affectionate nickname of Ferrari's late-1950s racers, called *Testa Rossa* (redhead) because of the red crinkle enamel on the cylinder heads. It, along with the Lamborghini Countach, is one of the most sought-after cars in the world.

The engines on the Testa Rossa, the Countach,

the Lotus, the Porsche 928S, and others of their rarefied ilk are the result of absolute refinement of all the foremost elements of automotive design over the years. Each one is superbly built, incorporating the latest advances in metallurgy and electronics, with extremely high compression ratios, multiple valves per cylinder driven by sophisticated trains, and every other innovation imaginable. Yet in truth, they are evolutionary engines—the product of years of experimentation along the road of high performance, with no concessions to economy.

Perhaps the most extraordinary aspect of any of these supertuned, superpowerful, superresponsive engines is the gulf between the power they can generate and the average driver's potential to use it. But this simply doesn't matter: if you have it and wish to flaunt it, you can do so better in one of these supercars than in anything else.

The engine specifications of the two Italian cars are very similar, at least in their Gargantuan proportions. The Testa Rossa has a double-overhead-camshaft flat twelve, with forty-eight valves; the engine delivers 380 horsepower at 5,750 rpm and can propel the two-seater to 60 mph in 5.3 seconds. The Countach has a V-12 capable of 420 horsepower; it, too, has forty-eight valves, and it features independent fuel injection for each bank of cylinders, enabling the car to reach 100 mph in 10.6 seconds, should you need it to do so. Although top speeds are disputed, either car can hurl you down the highway at 160 mph or more, if you dare. Economy-minded daredevils may be interested to know that the Countach costs about $9,000 more than the Testa Rossa's $109,700, and gets on average only 6 mpg to the latter's 10. Those fuel bills add up.

Cars of this type could be dismissed as extravagant expressions of conspicuous consumption—but so could diamonds, yachts, villas in Barbados, and weeks spent on Palm Springs fat farms. And unlike these extravagances, the great supercars have pioneered much in the way of automotive refinement that can now be found in standard plebeian production cars. But refinement is not confined to the engines of cars for the superrich. In response to a strange combination of competition, fuel shortage, auto magazines, and gradually increasing public awareness, engines far superior to any that existed in the past are now within the reach of most.

Powerplants for the '80s and '90s

The latest crop of automobiles, foreign and domestic, possess some of the finest engines ever built, from almost any standpoint except that of first cost. They are more fuel-efficient and more reliable; they demand less service; and they offer exciting performance despite being choked with complex emissions-control systems.

There is a price for all of this progress, of course, and we have learned to expect the $15,000 American family sedan; yet the press of competition is inexorable, and new marques from Korea and Eastern Europe startle the public with advertised prices of $4,995 (and selling prices that fall within perhaps $2,000 of that figure). A reasonable balance is emerging, as both Detroit and its old nemesis Japan adjust to these new price predators.

By and large, Americans have to thank the oil-exporting nations for the newer, more efficient engines. Had it not been for OPEC's fuel price manipulation, the classic monster American car would have been driven far into the future, with insensitivity to the waste and blissful unawareness that it handled like a spastic cement truck.

Who was to know? Generations had grown up with huge-displacement engines that were powerful and gas-guzzling. Before about 1939, they were considered excellent because in high gear they could lug folks along from almost a dead stop to normal running speeds. Later, the same rugged power was useful for overcoming the slushiness of early automatic transmissions.

In the past, there had certainly been examples of that radical idea, the economy car, in the form of the Willys, the Crosley, the Henry J, the Hudson Jet, the Rambler, and many others. The Volkswagen Bug, pounding happily along on its little flat-four air-cooled engine, had been embraced by millions. Detroit managed to ignore the VW with hauteur, only to find that it was just the stalking horse for the real competition.

The cars that came from the Orient didn't follow their American small-car predecessors in any way. Instead of being stripped-down, compressed big cars in the Henry J fashion, they were intelligently sized, well-equipped, and powered by hot little four-cylinder engines that got at least twice (and sometimes three times) as many miles per gallon as American cars. That was bad enough, but the Japanese went further and, while maintaining adequate interior space, pared down the bulk and weight of their cars so much that the resulting subcompacts delivered marvelous performance as well.

It was downright un-American—and even more un-Detroit. Much has been written about the aloof and arrogant management of Motor City, most cogently in David Halberstam's *The Reckoning,* but even these hidebound executives realized that changes had to be made, and the consumer has benefited. American cars are being upgraded in quality as they are downsized in bulk, and American engines have perforce adopted the attractive characteristics of their Asian and European competitors.

At the beginning of this chapter a (perhaps) invidious comparison was made between the simple, straightforward engines of the Model A and its contemporaries, and the almost occult clutter of a modern engine. The ultimate purpose of the illustration, however, was to show that you really don't need to know what all the new appurtenances are because they make the modern engine so much more reliable than its historical antecedents. The chances of your needing to get under the hood with a flashlight and screwdriver are very nearly as remote as the possibility that you could fix the problem anyway.

Upon analysis you can see the ways in which the automobile's racing heritage has devolved upon the modern engine. You can find a 2.0-liter four-cylinder engine with sixteen valves, double-overhead camshafts, and computer-controlled electronic fuel-injection powering midline four-doors with aplomb and panache. These superb little powerplants not only deliver more-than-adequate acceleration and top speed, they do so while providing power for air conditioning and a full deck of accessories. All they require in return is routine maintenance and regular oil changes. Similarly, advanced V-6 engines—the problem of balance solved—now offer admirable solutions to the problems of size, economy, power, and pep.

And although these may not be the engines that Henry Ford would have built, they are complemented in the marketplace with less exotic but still advanced four-cylinder engines of smaller displacement that power the Hyundais, Yugos, Horizons, and others. It is somehow fitting that the turbulent history of automobiles from the Model T through the golden age of tail fins is now resolving toward sensible cars available at relatively sensible prices. And what could be more appropriate than that the world's best-selling automobile for the last few years is once again a four-banger Ford. It's called the *Escort,* and in many ways it justifies the title of "world car" that has been applied to it.

TOP:
In 1953, the glamorous Packard Caribbean convertible relied entirely on the flowing lines of its body for its beauty, eschewing chrome strips, sculptures, or other excess ornamentation.
MIDDLE:
Chevrolet rocked the low-priced market when it introduced its 265-cubic-inch, 162-horsepower engine in 1955. This 1957 Bel Air, with its hard-top styling, became among the most-desired cars in America.
BOTTOM:
Ford, feeling the pressure from Chevrolet, restyled in 1955, and adopted a bigger engine that matched Chevy's in horsepower.

OVERLEAF:
In 1958, Ferrari took a giant step toward becoming the ultimate combination of styling and speed with the 250 Testa Rossa. One of the most successful competition cars of all time, its 300 horsepower V-12 was installed in a body by Scaglietti.

C H A P T E R 3

CARRYING THE LOAD

We are greatly beguiled by those thundering horses under the hood—all aspirated, turboed, hemisphered, quattroported, perhaps desmodromicized, or even given to injecting fuel. Nevertheless, the wild horsepower must be channeled into a suitable framework if it is to carry us over whatever roads we choose. Today, indeed, we even require them to carry us off-road in entirely new and more rugged vehicles.

Early Insights

At the advent of the automobile in 1895, manufacturers drew upon either carriage or bicycle practice for the creation of the chassis. Thus, the very earliest Benz used a chassis based upon bicycle experience, while Daimler selected a carriage type. And at a time when engines were capable of only 1 or 2 equivalent horsepower, either was satisfactory. Eventually, however, the demands of greater power and greater utility forced manufacturers to begin paying as much attention to the bearing elements as to the powering element.

Much as with the development of engines, parallel lines evolved, and among them certain benchmarks stand out. And again as with the engines, some very advanced ideas occurred to designers very early on. Chains were replaced by a shaft drive on a Renault in 1894. (Owners of the Oldsmobile Toronados that appeared in 1966 would probably be surprised to learn that their torque converter was connected to the gear box by a chain, for power transmission.) Four-wheel drive, so popular now, appeared on a 1900 Austrian Lohner, designed by none other than the genial Ferdinand Porsche. In the same year, the tubular steel frame was introduced by Darracq. The all-steel frame debuted on a Mercedes in 1900, and in 1902 the Humber featured an adjustable steering wheel. Disc brakes appeared in 1903 on the idiosyncratic but delightful Lanchester, and hydraulic brakes came on the English Hutton in the following year. Four-wheel mechanical brakes were used on the 1910 Isotta-Fraschini. A fully automatic transmission was available to official Mercedes state cars by 1914. (A very complicated electric automatic transmission was produced by Entz in the United States in 1898, and the same design was used by the later, better-known Owen Magnetic.) Duesenberg and Kenworthy both experimented with the first four-wheel hydraulics in 1920, and were followed later by Rickenbacker. Other manufacturers were cautious, introducing (as Chrysler did) hydraulics on the front wheels and mechanicals on the rear. Also in 1920, Alvis came

OPPOSITE:
The Buick was so famous for reliability that it became known as the "Doctor's car." Shown here is a 22-horsepower, side-entrance Tonneau.

BELOW:
Elements of both the carriage and the bicycle are evident in the Hammel car, circa 1880.

Buick

out with a synchromesh transmission that had a similarly gradual introduction. By 1922, in Italy, the Lancia had begun to use unit construction. Perhaps the most comprehensive bundle of these early insights was the luxurious Belgian Imperia of 1936, which offered a V-8 engine, front-wheel drive, and a fully automatic transmission.

But mere invention is not enough. Improvements have to be economical, producible, maintainable, reliable, and these logistical issues took time to resolve with respect to chassis, suspensions, and brakes just as they did with respect to engines. The various elements not only have to be improved themselves, they must be brought into consonance with other mechanical elements. A superb engine in a rugged frame is useless if the springs and suspension don't permit it to sweep swiftly around corners.

Assembly versus Full Manufacture

The improvement process was a long one, and the roads of the world were the test track. These conditions fostered the early and sustained concept of building "assembled cars," which did not receive the factory testing of modern automobiles.

There were perhaps few businesses in which investors saw more profit than the automobile trade during its infancy, and there were very few that engendered more pride. To name a car after yourself was a magnificent ego trip, particularly if you were trying to create a prestige vehicle. Even if not, it was exhilarating to be in on the ground floor of a new industry and to have a chance of becoming another Henry Ford. It was easy enough to try: arrange for a factory (anything from a local garage to an abandoned cannery), order parts from the plethora of manufacturers already in existence, and hire workers to bolt these parts into a finished car.

Of the more than 5,000 different makes of cars in U.S. automotive history, many followed this pattern, in which almost every major part was purchased outside and then assembled under the maker's marque. In many instances the body work

PAGES 90–91:
Considered by many to be the most beautiful production car ever built, when well and carefully maintained the 1937 Cord was an exciting car to drive as well. In 1937 the new head of Cord design, Alex Tremulis, added the brilliant touch of the chromium-plated supercharger exhaust pipes.

A racing Singer on the Avenue des Acacias, Paris, June 5, 1912. *(Photograph by J. H. Lartigue)*

(building wood frames over which sheet metal was spread) was done by the assembler, and everything else was acquired. Sometimes even the bodies were purchased. These makes often achieved a loyal following within a specific geographic area; a few became nationally known. Others were as ephemeral as will-o'-the-wisps—names that might be advertised once before disappearing forever (the disappearance often coinciding with the flight of the promoter with the stockholders' funds).

The generic term *assembled cars* was one of respect or derision, depending on the point of view. Car assemblers defended it on the basis of economics and quality control; full manufacturers attacked it on the same basis. Firms that assembled cars advertised their use of "the best of everything" from prestige companies. Thus their cars might have Continental engines, Brown-Lipe transmissions, and electrical parts from Delco—every part from the very best maker. For very upscale assembled cars, V-8 engines from Cadillac might be installed.

In some instances, absolutely identical cars were made—body style, engine, interior, and everything else the same, except the name plate on the grille and hubcaps—and sold under different names by totally unrelated companies, giving rise to the term *badge engineering*. The same concept was applied within larger companies, par-

In 1949 Nash introduced a bulbous body style that was called Airflyte by the manufacturer, but "bathtub" by many owners. This 1951 Ambassador Super featured, among other things, a reclining seat which could be made into a bed. It was popular with both salesmen and sportsmen.

ticularly during troubled times. Cars of one type would sometimes be given the name of another, more prosperous make, as a forlorn sales device. Thus we saw the last Hudsons standing embarrassedly in their Nash suits, and the proud name of Packard expire in unbefittingly sad Studebaker drag. Alternatively, junior editions were sometimes introduced with new names, as with Buick's Marquette or Cadillac's LaSalle, in the hope of skimming the marketplace.

Companies that made most or all of the parts for their automobiles extolled the virtues of their quality control and the economies of scale. Great firms such as Pierce-Arrow went to vastly uneconomical lengths to maintain control, maintaining (for example) exacting records of each batch of steel used in parts manufacture. Others, such as Franklin, continued to produce in-house standard parts that could have been purchased elsewhere, long after production rates had sunk too low for this to be practical. The public didn't care who made the generators, water pumps, junction boxes, or even transmissions, as long as they worked. Unless production was at so high a rate that internal manufacture of parts could be done economically, full manufacture was a luxury best forsworn.

The truth about the relative merit of assembled cars versus full-production cars not only lay very much in the middle, it changed over time. Economic necessity sometimes forced decisions that the

Although the rear view of the 1933 Pierce Silver Arrow has a Darth Vader quality, it is actually a very intelligent design solution to the problem of terminating the swooping lines of the side in an aesthetically pleasing manner. The "reverse eyebrow" windows were raised to give a modicum of visibility through the rear.

manufacturers might not have chosen to make. In 1954, for example, haughty Rolls-Royce adopted a Hydramatic transmission from General Motors because the cost of developing its own alternative was prohibitive and because such a unit was unlikely to be better than GM's.

Policies had to be altered to suit new conditions. Ford started out as an assembled car, pure and simple, with the Dodge brothers supplying enough engines to encourage their setting up their own make. Later, Ford attempted to integrate vertically, buying mines, forests, ships, and glass factories in an effort to control every stage of production of every component in the car. So canny was Ford that he used to specify how the boxes were to be made for materials that were shipped to him. When the boxes were disassembled, they could be used for floorboards and similar parts. If Ford had been Swift, he would have found a way to use the oink in the pig.

Ultimately all manufacturers used purchased parts wholeheartedly. It simply doesn't make sense to make everything in-house, especially items that a specialty manufacturer can update more easily. There is a hazard in this, of course: a major manufacturer could tie up huge sums of money in inventory, while at the same time placing itself at the mercy of the vendor. Japanese manufacturers solved both problems at once, however, by reducing the number of parts on hand to coincide almost exactly with production flow and by generating such competitive heat that suppliers dared not try to victimize them.

Most firms didn't publicize their practice of buying parts ready-made, even when the part came from a sister unit. In later years, when great overlapping was evident in the model lines of the big manufacturers, major elements—bodies, engines, transmissions, anything—were swapped among divisions to keep costs down. This should have been clear to anyone, with Mercury Bobcats looking like Ford Pintos, and Pontiac Tempests corresponding closely to Olds 85s. Yet so naive were American consumers, after more than half a century of intimacy with cars, that it came as a great surprise to many to find that some Oldsmobiles were powered by Chevrolet engines, or that some Dodges made extensive use of Plymouth components. The reaction was surprisingly vehement—more nearly resembling an adultery suit than a consumer complaint—and ultimately legislation required automakers to specify the origin of their components.

The modern buyer is more sophisticated and less blindly loyal to a brand name. We have completely adjusted to the idea that any given automobile may have an engine from Japan, a transmission from Germany, a design from Detroit, and assembly in Brazil. Even whole factories and their workforces may be leased, as in the case of Chrysler K cars built in the American Motors Kenosha plant.

The Backbone of the Automobile

The development of sophisticated chassis took rather longer than you might expect, primarily because expectations at the time were so low. People were used to the rather sharp, bumpy ride of the carriage, just as they were used to rutted and pitted roads. The speed of the fastest carriage did not require much in the way of banking on curves, and the type of springs that had been useful in the eighteenth century were not terribly out of place in the early twentieth.

But more powerful engines put far more stress on the frame, both directly from their mounts and indirectly from the beating they now forced it to absorb from the roads. As noted, Mercedes went to an all-steel frame in 1900, and in the next decade most manufacturers followed suit. A notable holdout was the Franklin, which used wood until 1927.

Wood, usually reinforced with metal gussets, had some positive values when used in the frame: it was resilient, and it helped dampen the shock of transit as well as the vibration of the engine. In time, however, most frames evolved into a sort of steel ladder, to which an engine was attached in the front and an axle was attached in the rear. The rungs of the ladder changed over time to K and X shapes and to variations thereon. Suspension systems were used both to attach the wheels to the frame and to isolate the frame from the bumps. Some brave pioneers mounted the axles over rather than under the frame, creating an underslung arrangement that lowered the center of gravity and (more importantly for sales) imparted a sleek and sporty look. Fred I. Tone's work, the American Underslung, is the most famous of these, although it is remembered more for the catch phrase than for the automobile, which was built in relatively small numbers for only eight years.

In general, most frames were overbuilt. The difference in cost of metal or fabrication in a heavy frame versus that in a light one was not significant, and using a basic heavy steel frame gave the manufacturer a great deal of leeway in the choice of engine installation and body styles. The frame could be built, the chassis completed, and the engine of choice installed, and then virtually any kind of body could be dropped on. Repair of both body and frame was also made easier.

In time, however, considerations of both strength and weight began to gain importance. As ever-heavier engines were installed and as bodies grew in size and weight, the frames had to be beefed up accordingly. The direct evidence of a need for increased frame strength was often manifested in the flexing of the bodies, with doors being forced open at speed. In cars equipped with aft-hinged "suicide doors," a sudden jolt could whip the doors apart. It provided the same sensation of imminent death that can be experienced when a jet plane's window suddenly opens next to you. The standard engineering solution of using more metal soon reached a point at which it intruded on performance. Palliative solutions were found in the suspension, as we shall see, but the direct response to the problem came with the integrated body/frame combination.

It is difficult for modern engineers familiar with the construction of ships and aircraft to understand why the unit body, which integrated the frame and the body into one structure, was so long delayed.

Calling them suicide doors was perhaps extreme; and there were often sound structural reasons for positioning the door at the rear, besides the obvious extra ease of entry they afforded. But suicide doors were found during the days before seat belts had been heard of—to have one pop open at speed was a never-to-be-forgotten experience.

The first reason was no doubt tradition: carriages had been built with separate frames, and so that's how most automobiles were built, too. Second, the need had not emerged immediately; and third, unit-body construction was inevitably more expensive than the traditional methods. The tradeoff had to come in performance—better acceleration and better mileage because of lighter weight and greater structural rigidity.

There were offsets to these advantages that had much to do with the level of resistance with which manufacturers greeted unit-body construction. It was an expensive process that had to be amortized over long production runs. Styling changes were difficult, and repair was often expensive. An unsuspected major flaw was corrosion. Standard chassis frames rarely melted from salt and road dirt, no matter how badly their bodies rusted. In contrast, an early unit-body car that had received rough treatment might sag in the middle, warning prospective used-car buyers to look elsewhere.

The first unit-body car was the 1922 Lancia Lamba, a remarkable vehicle in a number of ways. It had a very narrow V-4 engine, sort of a squeezed V with only a 13° space between the banks rather than the 90° space of the Ford or Cadillac. So narrow was this full-pressure-lubed engine that a single overhead camshaft could operate the valves of both banks. But it was the chassis that offered the greatest departure from standard practice. The entire frame and the lower half of the body were built as an integral unit, resulting in a strong, light structure. The design also featured independent front suspension and

four-wheel brakes—altogether a heady combination. From the model's introduction in 1922 until 1931, 13,000 Lambas were made.

Citroën produced its famous Traction Avant beginning in 1934. Over the next twenty years, 750,000 examples of the ''gangsters' car,'' with its pioneering front-wheel drive, would be built—its famous unit-body construction performed under license from the American Budd Corporation. Vauxhall built unit bodies from 1939 to 1948; and during the postwar period, such bodies would be adopted by many more manufacturers.

In America, Nash adopted unit construction for a mass market in 1941 and introduced coil springs all around, with sliding-pillar independent front suspension as seen on the Lancia. The Nash had been preceded in unit-body construction in the United States by the Cord and the Lincoln Zephyr, both appearing in 1936 but built in relatively small quantities. In most respects the Cord was the more technically advanced, for the Lincoln Zephyr used a controversial twelve-cylinder engine that had been cobbled up from a Ford V-8 heritage and almost incredibly stayed with the traditional Ford transverse leaf springs and solid axle arrangement. The Nash was in another league entirely: it was a typical mass-production car.

America Sets Its Own Course

Unit-body construction was gradually adopted by more and more manufacturers around the world, eventually becoming almost the standard technique. In a similar fashion, the chassis and drive trains of American cars were gradually improved. For years the most common practice had been to use solid axles front and rear, with semielliptic springs used to absorb the shock. Over time, rather dainty shock absorbers were added, but more frequently the remedy was to double up on the springs in size, length, or number of leaves. Many variations on the theme were undertaken, ranging from the Ford's traditional single willowy transverse springs to the Cord L-29's husky double-transverse springs (which provided a crude form

The front-wheel drive of the Cord was its most advanced and notorious feature. The car was fast, and its advertisements used to say, in effect, that you were passed in a Cord only when you permitted it.

of independent front suspension) to elaborate arrangements of full, half, or other combinations of leaf springs.

The ride achieved with these sorts of suspensions was adequate over smooth surfaces. Bumps tended to reinforce themselves, however, and the leaf spring was not good at negotiating rough roads at relatively high speeds, since its tendency was to jounce wheels off the ground, whereupon the driver lost steering control, traction, or both.

Suspension began to improve steadily after 1933. Maurice Olley designed an independent front suspension with upper and lower A arms linked by a coil spring, and this design appeared on the 1934 Cadillac, LaSalle, Buick, and Oldsmobile cars. A different system, from André Dubonnet, was optional on the Chevrolet and Pontiac. Both systems were called "Knee Action," but the massive shock absorber incorporated into the Dubonnet system made it prone to shimmy, so it ultimately was replaced by the Olley type.

Chrysler offered basically similar front-end systems in 1934, although the solid axle reappeared the following year. Ford stood apart, using solid axles and transverse springs of Model T mien through the 1948 model year. Independents followed GM's lead as well as they could—in this as in other things—and rear suspension almost uniformly consisted of twin longitudinal leaf springs, with minor variations in shock-absorber type and disposition.

Other elements, including semifloating rear

The Citroën is noted for having been one of the most advanced cars of its time in both engineering and styling. This model, more than twenty-five years old, is still remarkably modern looking.

axles, hydraulic brakes, free wheeling, synchromesh, and finally automatic transmissions, were introduced over the years in a like manner. Their introduction was always the product of a mystical amalgam of the need to reduce manufacturing costs by a penny or two per item and the desire to offer improvements that could be advertised. The self-destructiveness of this fundamentally incoherent process is ably recorded in David Halberstam's *The Reckoning*.

The European Divergence

In Europe, several factors combined to permit a greater variation in types. These factors included low production volume (which allowed customizing), a protective tariff (which excluded foreign competition), and a tendency to keep cars longer because they represented such a significant investment to European car buyers. A very successful make might persist for decades on an annual production of 7,000 to 8,000 cars; production of 35,000 cars a year signified astronomical success. People viewed their cars more as they might their home, not something to swap every three years, but something to cherish for its faults as well as its virtues.

The Lancia Lambda of 1922 was a true landmark in design, handsomely embracing a unitary body design with independent front-wheel suspension. It has a very unusual narrow-V, four-cylinder engine of 49 horsepower.

OVERLEAF:
The 1938 Jaguar SS-100 Coupe prototype was built for Sir William Lyons. The SS-100 series was intended to spur interest in the standard Swallow line, but instead it created an enormous demand for the car itself.

The weight of class in European society was very real. In America, the whole thrust of automobile advertising was a welcome assertion that the car defined the class: anyone could be upwardly auto-mobile simply by coming up with the money—a little down, a little per month—to drive away in the appropriate symbol. In Europe, if by some chance a clerk should come into possession of a Bentley, a Mercedes, or a Delahaye, he or she would undoubtedly continue to be perceived by all (including himself/herself), to be a clerk first, last, and always. The American illusion, pursued so assiduously by consumers and their suppliers, was different. Owning a Cadillac changed your status, no matter what else people knew about you and how well they knew it.

The difference in the European climate resulted in the flourishing of small-volume manufacturers that did not attempt to change body styles every year—or even every decade, in some instances—and could pursue engineering refinements on an individualized, sometimes idiosyncratic basis. Even though Detroit's emerging Big Three (General Motors, Ford, and Chrysler) operated subsidiaries in Europe, their products were treated rather off-handedly and did not gain a comparable ascendancy to the one they enjoyed in the United

States, through either volume or advertising.

Not that all American cars were bad, and all European cars good: there was plenty of room for variations in quality on both sides of the Atlantic. But the greater variety of developmental paths permitted by the European market enabled certain basic excellent mechanical elements to evolve there; and some of these were adopted in the United States only after the fuel shortage caught our attention. On the negative side, the broad spectrum of European marques offered specimens that bore many faults, including limited durability, difficult maintenance, sometimes abominable brakes and steering, cancerous rusting tendencies, and extraordinarily poor crash survivability. More important, the very things that permitted so many types of customized vehicles to develop worked against the type of general prosperity achievable only through mass production. Certainly, producing cars in quantity has been one of the secrets of Europe's financial resurgence in recent years.

Over time, some interesting divergences grew up within Europe. In England, poor roads and an antic method of computing horsepower based on the bore of the cylinder tended to produce cars with very small long-stroke engines and marvelous gear boxes to help them negotiate the curving, narrow lanes. In a few notable exceptions to this tendency, inspired engineers mated powerful engines to chassis that could handle them. The Vauxhall Prince Henry of 1914 was remarkable in this regard, while the original Bentleys were astounding cars, capable of winning at Le Mans against the best the world had to offer. Later glorious types include Aston Martin, Lagonda, and Frazer-Nash.

On the continent, the situation was far different. France had developed whole lines of magnificent marques—Bugatti, Delage, Lago-Talbot, Hotchkiss, Delahaye, Panhard-Levassor, and Bucciali—as well as some of only slightly lesser stature such as Peugeot, Renault, and Citroën. The Bugattis were perhaps a breed apart: a car for the very, very rich, infinitely complex, difficult to service, but always unforgettable to drive. In

Wales, August 1924. (In 1951, the author experienced the same degree of surprise when the same thing happened to him during what turned out to be his last drive in a faithful 1932 Buick Victoria coupe.)

OVERLEAF:
This 1935 Hispano-Suiza K6 coupe, with its flamboyant body by Saoutchik, was the distinguished French firm's attempt to develop a smaller alternative to its twelve-cylinder engine. **The K6 basically had half of the bigger engine, and was offered with power brakes and softer steering. As a result, the chauvinists of the day referred to this behemoth as "The Woman's Hisso."**

10
CT 6559

comparison, the others—while hardly Model Ts—were reasonably inexpensive to operate and (with a modicum of skill) could be maintained by the owner.

All of these cars featured excellent suspension systems suitable for high-speed cruising on the fine French roads. In such sportif cruising, of the sort Hemingway describes vividly in *The Sun Also Rises,* the road hazards were not other cars but herds of sheep or cattle.

In Germany, similar cars were produced in relatively small numbers but were available in a surprisingly wide choice of makes. Mercedes, with a nearly perfect heritage in terms of invention, established its preeminence early through success in courting a prestigious clientele and success in withstanding the hard knocks of racing. (American chauvinists are always pleased to learn that, from 1910 to 1923, Mercedes manufactured under license the Knight sleeve-valve engine type that graced the American Willys.)

Mercedes attended (in relatively small numbers) to the needs of the middle class by offering the 170 series, introduced in 1931 and continued in largely unaltered form into the postwar years. In the beginning, it featured a rather small 32-horsepower engine, rubber-mounted to the chassis. More important for the future was the

With today's proliferation of cars from every country, it is difficult to remember the visceral impact that the Jaguar XK-120 had upon the automotive world at its debut in the London Motor Show of 1948. It had everything—beauty, style, engineering, and a fabulous engine. In 6 years, only 12,055 would be built, but they would never be forgotten.

adoption of independent front and rear suspension. Power was increased over time, along with other improvements; and despite the interruption of the war, almost 90,000 representatives of the 170 series were produced between its debut year and 1947.

But the real Mercedes image is built upon the far smaller numbers of classic road cars that are so highly prized today. These are epitomized perhaps by the 1934 500K Roadster, with all-independent suspension, an eight-cylinder supercharged engine, and the sort of sound and look that turns car-lovers' insides into jellied consommé. Mercedes models were always heavy, requiring not only big engines but the occasional boost of a supercharger. The latter could be operated for only a very few seconds at a stretch before causing tremendous induction problems, so it must be viewed as more of a kick-down accelerator than a full-fledged power augmenter.

Mercedes was not alone. Other firms turned out enormously successful cars in very small numbers. The most enormous (and the rarest) of these were the Maybach Zeppelins. Fantastically expensive (as much as $40,000 in 1937 dollars), the Zeppelins featured great twelve-cylinder engines, sophisticated suspension, and elaborate gear boxes equipped with eight forward and four reverse speeds. The Germans had a word for such a transmission, of course: *Doppelschnellgang,* or double overdrive.

Another magnificent marque was the Horch, one of the antecedents of today's Audi. Its appearance was more subdued than that of the Mercedes, but its engineering was quite as sophisticated. The Horch was a highway cruiser, offering all the luxury of the Mercedes but lacking its extremes of performance.

In 1936 BMW produced one of the great sports cars of the age with its 328, which sported an elongated version of the famous trademark kidney-shaped grille. With an 80-horsepower, six-cylinder engine, a tubular chassis, and lovely flowing lines, it used rack-and-pinion steering and independent front suspension. Only 462 of these lovely vehicles were built, but they had an influence all out of proportion to their numbers in both Germany and Great Britain.

Other famous names emerged elsewhere. Isotta-Fraschini established Italy's claim to elegance, just as Hispano-Suiza did for France and Spain. In Belgium, the Minerva and the Imperia were classic makes—national equivalents of the Rolls-Royce or the Duesenberg.

But circumscribing these successes was the matter of scale. In 1937, England's production was the highest in all of Europe, peaking at almost 380,000 cars. In the same year Italy built 61,000, Germany about 110,000, and France about 190,000. These totals were shared among a multiplicity of manufacturers, and they compare with the figure of more than 3,000,000 cars built in the United States.

The war, of course, devastated the production and design of cars, but the postwar recovery was boosted by a new and different appreciation of the automobile in Europe. After so many years of deprivation and in the face of so much technological development, Europe became a hothouse of engineering and styling refinement.

Perhaps the most influential designs came from England, where two cars grew to greatness inadvertently. The luxurious cruisers we now know as Jaguars began their lives as sporting spinoffs of a firm called SS Cars, Ltd, at a time when SS did not have its present dreadful connotation. When the Jaguar XK-120 burst upon the scene in 1948, the automotive world stood still. Intended as a loss leader to pump sales of Jaguar sedans, the XK-120 was met with overwhelming demand. Here was a car with a beautiful body, a magnificent engine, and an excellent road-racing-developed chassis, delivered at a price that was very competitive, about $5,000.

Supporting it ably in revolutionizing the motoring psyche was the M.G. TC. Available for half the price of the XK-120—with about two-thirds the performance, and one-fifth the comfort—the M.G. enchanted Americans. Phrases like "four-wheel drift" and "he got me on the straights, but I took him on the turns" began to be heard at tony American country clubs.

It was absurd. These cars carried two people in cramped discomfort; their tops leaked, their heaters rarely worked, and their rough suspension systems let you feel every tar-joint in the road. Yet America went bonkers over them. Perhaps even more remarkable was that this enormous effect was attained from production runs that Detroit would have considered not only trivial but terminal for an American car. There were 12,055 Jaguar XK-120s built between 1948 and 1954, and 10,000 M.G. TCs built between 1945 and 1949. Altogether, that's 22,055 cars—not all of which were sent to the U.S.—responsible for initiating a revolution in a country that produced 40 million cars in the same time frame. It should have served as a signal to someone other than George Romney and the Japanese.

Perhaps Detroit might be forgiven for dismissing this English incursion as a temporary fad, but there was no such excuse for ignoring the lesson of the Volkswagen. This charming car rapidly achieved and tenaciously maintained cult status in the United States, for every correct reason. It was small, economical, peppy, and distinctive. Just as Perrier gave people an excuse not to drink hard liquor, so did the Volkswagen give people an excuse not to drive a relatively expensive, boring car.

The roots of the VW run back to 1932, when Ferdinand Porsche created a rather large 3.5-litre car for his personal use. Called *the Wanderer,* it had the now-familiar curved back outline. He pursued the same line of design in building a smaller car powered by a rear-mounted 1,200-cc engine and equipped with other VW portents such as a swing-axle rear end, front-transverse leaf suspension, and integral headlights.

Hitler saw the need of a *Volkswagen* (people's car) to provide a plausible cover for the essentially military motive underlying his plan to build a network of autobahns throughout Germany. He was a car buff of the first order and an admirer of Henry Ford. A few months after he came to power in 1933, he met with Porsche and set down specifications for a car that would cost no more than 1,000 reichsmarks (about $430 in 1933 dollars), have a top speed of 100 km per hour, carry two parents and three children (a modest request, given the Nazi emphasis on fecundity) and get better than 40 miles per gallon in gas mileage.

A band of Volkswagen Bugs. Designed by Dr. Ferdinand Porsche, the VW ultimately became the right car at the right time, and more than twenty million were sold. *(Photograph by R. Lessmann)*

OPPOSITE:
It is difficult to connect the malevolence of Hitler with the amiable, almost Disney-like disposition of the Bug, but the dictator had a profound influence on the conception and initial manufacture of the Volkswagen.

Porsche doubted whether these goals could be achieved, but with government financing he proceeded. Hitler even sketched a picture of what he thought the car should look like, remarking that the beetle was a perfect example of streamlining. The passage from drawing board to autobahn was carefully orchestrated. Eight prototypes were subjected to a 28,967-km test program over all parts of Germany, and the defects inevitable in any new design were discovered and corrected.

In 1937, the Volkswagen became a state-funded project. Thirty more prototypes were built, and these were subjected to a similar intensive test by members of the SS. The car was refined to include a rear window, running boards, and a slightly more powerful engine.

In May 1938, the foundation stone for a new factory to build the KdF-wagen (for *Kraft durch Freude,* or strength through joy) was laid by Hitler personally. That August, a unique payment plan was revealed by Robert Ley, chief of the Labor Front: the buyer could simply make payments totaling 1,250 reichsmarks, and when paid up could travel to the factory to pick up a brand-new car.

Orders poured in, ultimately reaching almost 350,000. Unfortunately the war intervened, and almost all who had faithfully made their payments never received a car. Less than 650 were built during the war, and these were reserved for the use of high-ranking officials.

After the war, the plant came under British control. Both British and American automobile manufacturers discounted the VW, failing to see the remarkably advanced ideas in Porsche's design. It was a case of rampant, infectious industrial myopia, in which an insultingly half-baked response to rising public interest was cloaked in the tinny sheet metal of Vegas and Pintos.

Heinz Nordoff became plant manager after

DEC CALIFORNIA
46 VW

1313

Volkswagen reverted to German control following the postwar occupation, and he proceeded to make the Bug the dominant car, first in Europe and then (in its class) in the United States. It became the central element in the German economic miracle.

The Volkswagen was brought to the United States for sale in 1949 but generated little interest. In 1952, however, 601 were sold; and by the year 1968, Americans were purchasing 423,008 Bugs annually, beguiled at least in part by the masterful VW advertising campaigns.

The Bug, with its familiar-sounding four-cylinder engine, might have gone on selling forever in the United States, but its design was not adaptable to the stringent emissions-control requirements established by the EPA, and as a result its manufacture was discontinued in Germany. Subsidiary plants in Mexico, Brazil, and Australia continued building the Bug, however; and VW remanufacturing plants are also springing up, where a loved Bug can be brought for a complete ground-up restoration that will make it good for another 100,000 miles.

The European conditions of multiple manufacturers and small anticipated sales led ultimately to the same sort of proud-name-swallowing consolidations that were taking place in English aviation and in the American automobile industry. But before the consolidations took place, various engineering, styling, and mechanical innovations were developed that were seized—first by the Japanese, and later by everyone—as a basis for creating the far better (if far more expensive) cars we drive today.

The first, most sensible, and most beneficial reaction was a general reduction in size and weight. (History seems to be repeating itself, however: after all the efforts to downsize and after the smashing debut of diminutive Japanese cars, both size and horsepower seem to be increasing in all cars of all countries.)

The second important factor was the obtaining of more and cleaner horsepower from engines of smaller displacement and higher rpm.

The third element, perhaps the one most appreciated by drivers, was the improvement in suspension, braking, and tires.

Coil-spring independent front suspension had been around for many years, but the introduction of such elements as torsion bars, McPherson-type struts of telescoping hydraulic dampers enclosed in coil springs, and very sophisticated geometries of suspension elements provide racing-car-quality front ends to family sedans. In a similar way, the solid rear end (except in the case of beam axles on front-wheel-drive cars) has given way to De Dion suspensions (a curved tube designed not to transmit power but only to carry the weight of the car), live rear axles, and other systems cradled with torsion bars, trailing arms, and superior shock absorbers.

These elements have all been calibrated to work together, as well—a process only possible through the use of extensive testing equipment that was not available even to the great manufacturers two decades ago. The result of this extraordinary cumulative effort is superb-riding cars that are capable of handling both powerful engines and curving roads with safety.

Disc brakes—in which circular pads are forced against the brake disc (in contrast, in drum brakes, "shoes" are forced against a curved internal surface)—first appeared on the 1903 Lanchester, but they were not introduced into production until they appeared (oddly enough) in the tiny American Crosley of 1949. Chrysler added more convincing support to the brakes with its adoption in the same year, and Jaguar soon followed suit, stamping the newly revived device with racing approval. But the real reason for the subsequent nearly universal adoption of disc brakes lies in the continued reduction in wheel size. As manufacturers sought to lower the overall height of their cars and (later) as the size of the automobile itself decreased, wheels were progressively reduced from an average size of 16 or 15 inches in diameter down to as little as 10 inches in diameter on the Morris Minor. These small wheels left no run for drum brakes, so discs had to come; and once adopted, they proved to

OPPOSITE:
Hitler and Henry Ford fortunately had little in common, but the Volkswagen production line was as much of a miracle for post-World War II Germany as the Model T had been for post-World War I America.

OVERLEAF:
The French understood automobile styling the way they comprehended Impressionism, champagne, and literature. The famous Parisian coachbuilder Saoutchik built this rollicking, joyous creation on the first Talbot-Lago Type 26 Grand Sport to roll off the production line after World War II. In many ways it paid tribute to the past by making a promise to a racing future.

be far superior to drum brakes in both stopping power and resistance to fading from heat buildup.

The most significant improvement in automotive braking has an aircraft history dating back to 1947, when an antiskid system (essentially a pressure-sensing device) was installed on the Boeing B-47 bomber. In a conventional braking system, standing on the brake pedal during a panic stop is apt to produce out-of-control skidding that uses up more distance than exists between the car and the object it unwillingly approaches. The new antilock braking system (ABS) in automobiles employs a computer to determine just when a wheel is about to lock up, and thereupon automatically releases the brake pressure just enough to prevent it. The resulting capacity for skid avoidance permits far shorter stopping distances for unskilled brake trompers—a title most of us have earned at one time or another.

The decline in wheel size, with its attendant increased burden on handling, has been more than offset by the vast improvement in tires. Americans had for years bovinely accepted the fact that great whitewall bias-ply tires would wear out in perhaps 15,000 miles. In 1949 the French company Michelin created radial tires. It took us time, but eventually we found that these tires could last 40,000 to 60,000 miles, while simultaneously providing far superior road-holding capabilities. The result was a virtual restructuring of the United States tire industry (which nonetheless fought off the obvious for as long as it could, before conforming).

All of these improvements have one thing in common: they are expensive. Simply put, the world was building better cars, and they cost more. For a while, sticker shock replaced heart disease as the number one killer in America—well, almost. But

RIGHT:
The Chrysler Airflow was a triumph of engineering over styling—almost always a Pyrrhic victory in the automotive world. An excellent car, with a good engine and running gear, comfortable to drive, it was, not to put too fine a point on it, ugly.

BELOW:
The M.G. TC taking the turns! This gravelly little rattler, of which a mere 10,000 were built, inflamed the desires of men and women the world over for a sportscar. It is not unfair to say that every Corvette, 300Z, Supra, Camaro, Firebird, Fiero, or Thunderbird on the road today owes a debt of gratitude to the M.G. TC and the Jaguar XK-120.

The Austin 7 survived throughout its long production period, not by providing the verve and thrill of an M.G. or a Jag, but by being a sensible car sensible people could drive in the country on the weekend without going broke.

Americans got used to it and now routinely think in terms of $15,000 family automobiles, while Mercedes, Porsche, Jaguar, and other luxury-car manufacturers have no problem selling all the cars they can make at prices that range from $40,000 to $75,000.

Among the thousands of makes of cars, perhaps three can be selected to illustrate the most significant turning points of the last thirty years in automotive engineering. The first was the advent in 1959 of Alec Issigonis' Austin Mini. (Issigonis had previously designed the Morris Minor, a handsome little car that was the first in Britain to exceed the 1 million mark in sales.)

The abortive Suez War of 1956 had given Great Britain the same sort of fright that the United States endured during the 1973 fuel crisis. Issigonis was assigned by Austin to build a car to meet the crisis. The result was the Mini, which placed the gearbox for the front-wheel-drive unit under the transversely mounted power unit, thus making 80 percent of the vehicle's space available for the passengers. It was a once-in-a-lifetime stroke of genius. This tiny vehicle, 3 m/10 feet long, established the format for the next generation of cars of almost all sizes throughout the world. As part of the general corporate consolidation characteristic of the period, Morris built the same car and marketed it as a Mini-Minor.

The second epoch-making car was the Datsun 240Z, which in 1969 heralded Japan's coming of age in the automotive world. It was sensational

BUICK
ROAD

The gap-toothed grin of the 1948 Buick Roadmaster.

when it appeared: a 122-mph car with a single-overhead-cam engine, dazzling looks, and a $3,500 price tag. Immediately selected for all of the Ten Best Car lists, the 240Z at a single stroke freed Japan from the pejorative econobox image that Detroit had encouraged to stigmatize her cars, and catapulted Datsun (now Nissan) into a leadership position.

The third car will surprise some readers, for it is the latest Toyota: the Camry. Here is a car that has melded the best advances of the last twenty years into a supremely smooth, amazingly reliable package. As much pleasure attaches to the fact that the Camry always works as does to its rather simple, clean lines. The results are evident in owner loyalty, general popularity, consumer reports, and (in the final determinant) sales. You'll note that none of the three cars mentioned was American-made; the reason for this lies in the longstanding dichotomy between American and foreign sales and manufacturing philosophies, which only began to come together under the onslaught from Japan. Perhaps the best way to illustrate the problem is to consider a semiunitary design, Chrysler Corporation's Airflow of 1934—a car that was clearly far ahead of its time in almost every respect.

A young engineer named Carl Breer recognized the traditional relationship of automobile to aircraft but carried it several steps further. He persuaded the Chrysler Corporation to pursue a development path using wind-tunnels, aeronautical engineers, many full-size prototypes, and genuinely pioneering thinking to create a totally new, more efficient car. In the process, he and his colleagues selected a semiunitary construction technique. A massive effort resulted in the 1934 Chrysler and De Soto Airflows. Streamlined, fuel-efficient, fast, comfortable, and competitively priced, they were unfortunately regarded as ugly by a public attuned to long hoods, sweeping fenders, and grinning grilles.

The resulting sales and corporate catastrophe supplied an American synonym for failure that endured until the arrival of the Edsel. It was a crystallization of the clear signal that American buyers had (perhaps without realizing it) been sending all along. The American public insisted with every ounce of its buying strength that it wanted the General Motors styling department to tell it how cars should look, *every year.* Like a desperate, overrun infantry company, the car buyers of America called in the artillery of planned obsolescence to fire on their own position.

Competing car manufacturers responded as well as they could to this vote of adoration for GM, relying as much as possible on the countervailing force of the historic "owner loyalty" syndrome. Ford had more loyal followers than Nash, Nash had more than Hupp, Hupp had more than Reo, and so on. Each time the buyer was wooed from an independent to one of the Big Three, another nail was driven into its coffin. The companies that possessed the resources to produce cars in the millions or hundreds of thousands annually were thus forcibly wedded to the task of building cars we thought looked good and performed satisfactorily within the parameters of our highway system. The manufacturers who built cars in the tens of thousands hung on by their nails until customer loyalty was at last eroded.

Production figures illustrate the process. In 1959, Studebaker's convulsive last effort, the Lark, spurred sales to a healthy high production of 153,823 cars; thereafter the production curve followed a descending line to oblivion, with 105,902 in 1960, 86,974 in 1962, and 67,918 in 1963. Exact figures are unavailable, but only about 10,000 Studebakers were produced in the last, pathetic year of 1966. Similarly, Kaiser-Frazer went from its 1948 high of 181,316 down a failing track to 16,759 in 1954, before folding after about 1,000 cars in 1955—most of the last year's production being shipped to South America. Pick any late lamented name—Hudson, Nash, Packard, De Soto, Willys—and the process holds true. America demanded both styling conformity and continuous styling change, and this could be achieved only by companies with big production volumes. The vehicle underpinnings of construction, suspension, and the rest mattered far less. The full implications —good and bad—of the styling elements of this planned obsolescence will be dealt with later, but

ABOVE AND OPPOSITE:
Cadillac sought to provide immediate product identification for its owners by endowing the remarkable 1948 series with a little bump of a tail fin, a gentle upsweep in which the taillight glowed. Within a few years, *everyone* was doing tail fins. *(Photographs by Curtice Taylor)*

the practical effect was to wed American technology to a course of separate body and frame construction, producing shells into which big engines could be mounted at the appropriate moment. The power from these engines was transmitted by conventional means to totally conventional rear ends. It followed, as night follows day, that the springing systems and shock absorbers chosen would be those adequate to contain the bulk of a car wallowing, heaving, and lurching forward at about 60 mph on standard American highways. This sufficed, so long as the shell of the car was graced with flow-through fenders, gap-toothed grilles, and tail fins or ventiports.

Upon analysis, the stylists seem to have suffered drafting-paper pituitary problems as they used their public mandate to make cars progressively bigger and heavier, reaching grotesque proportions by 1967. This striving for the behemoth resulted in some unfortunate driving problems: parking required either an 800-pound gorilla at the wheel or power steering; and stopping fostered first power and then disc brakes. In the process, amenities such as air conditioning, powered windows, powered ash trays, and other devices were ladled on with a gravy salesman's generous hand. To offset this, engineers were given the task of making driving easier (*easier,* not *safer*: at this point, the word *safety* ranked with *Airflow* and *Edsel*). Easing the task meant providing more power, and the size of engines grew, contributing their own bit to the problems of weight and size. The ultimate result was the Pleistocene American car of the 1960s and 1970s.

How could it have happened? It seemed easy at the time—as explicable as falling in love with a person who obviously buys all the right products. Let's go back in time and see what was considered beautiful, and see if we can determine what the elements are that made (and make) styling so important. It will be interesting to see if our analysis can explain why first General Motors and then Mercedes were able to clamp upon this infinitely varied world a set of styling mandates that all other manufacturers must slavishly follow.

C H A P T E R 4

AUTOMOTIVE STYLING: FORM FOLLOWS FUNCTION FOLLOWS FORM

A good engineer or a good machinist is inevitably a good designer. The simplicity, economy, and elegance demanded by an optimized working part also serve to create beauty, even if inadvertently. Illustrative case studies of the intertwining of art and engineering readily spring to mind—the most obvious being the resemblance between Brancusi's celebrated and initially controversial "Bird in Space" and a propeller blade. Extending our scope further, we can compare a glistening, functional Gnome-Rhone rotary engine and a Calder mobile; in both, moving parts that are essential to the whole nonetheless seem to move independently. The same measured mathematical fluidity is conveyed by the profile of a camshaft lobe as eloquently as by an Eames chair. And the comparisons can be extended into the computer age by examining both the little striped information block that causes a checkout stand's cash register to go "boink" and the vertical mysteries of a Gene Davis canvas.

Thus it was that the earliest cars of the Daimler and Benz era had a stark beauty of their own: a contained, elemental truth. The functions of engine and gears were not disguised or hidden, and the simple, functional strength and purpose that can be read in each practical line make these cars lovely to look at even today.

OPPOSITE:
Haute couture—and the consciousness thereof—have traditionally been closely associated with the most fashionable of automobiles. Here, in 1953 Paris, the elegance of a Cadillac convertible is paired with the chic styling of a Jacques Fath silhouette. ***(Photograph by Willy Maywald)***

It was not long before these native mechanical masterpieces were tarted up, however, for the automobile is more vulnerable than any other machine to the human touch. By way of contrast, consider that railroad trains were left as unstreamlined as a space station for more than a century, before changing public expectations led Raymond Loewy and others to bolt semismooth sheet metal to the standard bulbous externals of veteran engines. Like automobiles, aircraft went through a baroque period when simple embellishment and overstrength bracing were common, until they were foreclosed upon by the genuine need for streamlining. And the process continues on today's highways; now the huge eighteen-wheelers have drag-reducing shields mounted on the cabs.

In the long progression from 1895 until today, a few cars have been works of art, more have been beautiful, and many have been simply functional and, as such, attractive. In bell-curve fashion, we find on the other side of the aesthetic coin some cars that were pure junk, more that were ugly, and many that were simply functional and, as such, unattractive.

This distribution has prevailed over the span of a century and can be divided historically into five rough epochs. The first, from 1895 to 1915, we may term *Primitive,* in the affectionate and admir-

The almost immediate adoption of the automobile by heads of state gave a legitimacy to the expensive car that might otherwise have been difficult to attain. England's beloved Edward VII was well-known for his enthusiasm for the automobile; shown here *(left)* is his 1900 Daimler being driven to a 1930 auto show.

OPPOSITE:

The Cunningham was one of the best, most expensive, most handsome, and least known of American automobiles. Cunningham built huge vehicles with enormous wheelbases, which were the favorites of cognoscenti like Mary Pickford, William Randolph Hearst, Fatty Arbuckle, among other such luminaries who wanted to set themselves apart from the pack. This car is believed to be a 1919 DePalma Speedster.

ing sense that some pictorial American art is called primitive. From 1916 to 1926, the need for production standardization dominated, with the styling efforts subordinated to the process of learning how to produce complex cars by the tens of thousands. Here the term *Rational* might be useful. It should be noted that, for some cars—the Reo, the Locomobile, and others—rational decisions that the current product was adequate and that styling changes were not needed resulted in their disappearance from the marketplace. For others (most notably in the case of Erret Lobban Cord's legerdemain with Auburn), styling changes saved the company.

From 1927 to 1942, designers were often given a free hand, and while most cars were still produced from rational considerations, others were extravagant enough to earn the period the title *Baroque*. This is not altogether fair, for while virtuoso coach builders often embellished their great machines with curves of metal, wood, and fabric to a degree of opulence beyond average tastes, many drew their inspiration from the most valid functionalist principles and produced bodies of sculptural beauty. But it was a time of fun, when the advertising renditions in magazines depicted cars that would have required humans to have 5-foot-long torsos and 6-inch legs in order to be comfortable. The designers wanted cars to be long, low, and flowing, even if the roads and the chassis required them to be high, short, and dumpy.

After the hunger for cars created by World War II had been partially appeased, the period of the bombastmobile began, so ably characterized by Lichty in his *Grin and Bear It* cartoons of "Belchfire V-8s." It was a time of happy, egregious excess in which most people drew solace from acres of nonfunctional, rust-loving sheet metal—especially if there was more than ample chrome. The cars' generally improved performance capabilities were overwhelmed by the fixation on the outward appearance of the body work, just as a tender piece of veal can be submerged in an inept Marsala sauce. Let's call the period from 1946 to 1976 *Escapist*, for the process may have stemmed from an urgent need for relief after stomaching the most terrible historical sandwich of all times: a depression between two slices of world war.

The western world may not think it has much to thank OPEC for, but it did save us from the Escapist period, thrusting us today into a marvelous time when the demand for fuel economy has resulted in beautiful, efficient cars that offer superb performance. For the first time in history, mass-produced cars are attaining quality levels superior to those achieved by the great luxury marques—Rolls-Royce, Isotta-Fraschini—of the past. And the promise for the future is even more impressive. Perhaps the best name for the period from 1977 on might be the *Relearning* epoch, for it is in this time span that we have discovered anew the art of economy.

It should be borne in mind that, although some car names remained the same for a half century or more, their product lines and thus their image often changed drastically over the years. In discussing the various paths that styling took, we will find it necessary to recall periodically what the car was at the moment in question, rather than what it became subsequently, or how it is remembered now. More poignantly, we will be discussing some names that are not remembered at all—Peerless, Cunningham, McFarlan, Lanchester, Lohner, Sizaire et Naudin—but that once conferred the imprimatur of cultivated taste and old money.

C3-72-68

The Primitive Period (1895–1915)

The pioneers who were so immersed in the creation of the automobile would doubtless be offended by the term *Primitive,* but it is not intended to denigrate their efforts. Rather, the term is a relative one, focusing on the comparative effort expended on the task of styling versus the task of making an automobile.

The first step beyond mere motorized platforms was the Viktoria, introduced by Karl Benz in 1892. It marked the commencement of a long period during which adapting standard horse-drawn coach configurations to automotive design was seen as the primary styling task. The results stayed with us long after the automobile acquired an identity of its own. The name, if only a vaguely reminiscent style, persisted in such body types as limousine, victoria, sedanca, and brougham. Eventually, no sensible relationship remained between the actual body style and the name, but the manufacturers were nonetheless reluctant to part with the evocative nostalgia of tradition, so they applied the terms instead to individual series of cars within a make. Thus in later years we have Ford Victorias, Cadillac Coupe de Villes, and Pontiac Grand Ville Broughams whose names tell us nothing about the shape of the sheet metal enclosing the mechanical parts. Many people find this harmless affectation preferable to some of the coined names of recent years, which seemingly are designed more to achieve patentability than to evoke an image. Such legal considerations have given us Celica and Acura from the East and Toronado and Camaro from the West.

In the first decade of the twentieth century, numerous styles had evolved. The most common of these consisted of a square-rigged frame, wheels disposed at the farthest point at each end and covered by flat shield fenders, an enclosed engine space, and two seats carrying two or three people each. The steering wheel was tall and unmistakable, and various levers for clutch, gears, and brakes were disported almost anywhere the chassis designer chose—outside the body or inside it. Two 1909 types, polar in cost but of basically similar outline, were archetypal: the Rolls-Royce Silver Ghost and the Model T tourer.

Advertising art and automotive legerdemain merged perfectly with the Jordan Playboy. This is the same 1923 model that inspired Ned Jordan's "Somewhere West of Laramie" advertisement copy that set the ad industry on fire. For the first time in automotive history, the selling of a car was left to the car's inherent romance and panache, not its reliability or performance.

Some automotive designers were as precocious in styling as some engineers were in mechanics. The utterly handsome curved-dash Olds of 1901 was an amazing expression of beauty in a mechanically uncomplex automobile. Only a year later, one of the first attempts at drag reduction occurred with the aerodynamic Serpollet Easter Egg, a steamer capable of 75 mph. Renault created in 1900 a fully closed body on the Type B, attractively styled in a double-phone-booth motif. Powered by a rip-roaring 2.75-horsepower engine, the Type B featured a pig-snout hood nestled against the square body. More significant than either the car's engine or its body, however, was Renault's introduction of a three-speed gearbox with direct-drive power transmission.

The period saw endless styling variations, ranging from tiny voiturettes—scarcely more than powered roller skates—such as the 1907 De Dion Model AV, the 1905 Italian O.T.A.V., and the 1909 Le Zebre, to the luxurious 1910 Benz Phaeton. The shared feature of almost all the styles is the unapologetically peremptory manner in which they were obviously added onto the mechanical components. The bodies showed few curves and even fewer compound curves. More disconcerting was the prevalent discontinuity between elements of the vehicle. It was as if the designer said, "Well, here's an engine. Let's slap a box around it." Then, looking up, he might have commented, "A passenger compartment, eh? Let's slap a box around that front seat; and later, if we get to it, we'll put another box around the rear."

This multiple-box approach is evident in the Lanchesters, which were so eccentric as to have had a beauty all their own; in the Locomobile, which dared anyone to criticize; in the Peerless; and even in the immortal Packard. The Packard stylists, more than any others, were able to confer upon basically square lines a panache missing in

lesser cars. There was a crispness, a cohesion to Packard styling that made it an instantly identifiable yet unintimidating luxury car.

Most cars were open, with either no top at all or an awkward and drafty canvas folding top. In instances where a fixed top was required, however, such as in the 1909 Packard Model Thirty demi-limousine or the 1910 Peerless limousine, the results are quite harmonious.

The ad hoc styling attempts of the period can be traced to a number of considerations. The first was sheer novelty: *any* automobile was sensational, just by existing. Second, small numbers produced meant small factories, with few specialists. Third, and perhaps most potently, the hothouse flowering of engineering improvements diverted attention and resources away from styling considerations.

THE ARTIST INTRUDES

The beginning of mass production of automobiles with the 1909 Ford Model T introduced a new factor. Alert manufacturers saw that if they could not produce a car with the Ford's low cost, they had to distinguish it otherwise—mechanically and in appearance—so that the customer would be willing to pay the higher price. Mechanical refinements were more expensive to develop and had to be explained and demonstrated, while styling had the great virtue of selling itself.

By 1912, various cars had appeared that gave styling considerations equal importance to engineering matters. The Italian flair showed early in the sleek 1912 Zust touring car, whose inverted horseshoe radiator was neatly faired into a long engine cowl. As would so often be the case throughout the history of the automobile, smaller cars both imitated and innovated. The 1912 Singer 10 was a miniature version of the Zust, though the styling was certainly just coincidental; it was also far longer-lived, being in production for 27 years.

The contemporaneous French Gregoire, however, was clearly inspired by the 60-horsepower Mercedes produced from 1902 to 1905. And as frequently happened, the imitation was better executed than the original: smoother and more elegantly done. The Gregoire's hood and coaming were integrated, while the twin seats were backed up with a circular gas tank, reminiscent of those on the immortal American Stutz Bearcat and Mercer Raceabout.

The flurry of styling improvements that occurred toward the end of the Primitive period foreshadowed the future in more ways than one. The emphasis on sheer good looks was obtained by an economy of line and metal. These truly were cars that illustrated how "less is more" by avoiding the complex shapes of the past—where seats appeared to be copied from a Roman stadium, and hoods were buttresses flying against the square front end of the passenger section. In later years, the Ruxton, the Citroën Traction Avant, the Cord 810 (by coincidence, all front-wheel-drive cars), the Jaguar XK-120, and the Ford Mustang all similarly had beauty conferred upon them by the need to economize on design.

OPPOSITE:
By 1895, when this Panhard-Levassor *(top)* was built, the pioneering firm had built over 400 cars, and the public was clamoring for more. Another success, almost from its start in 1896, was the Delahaye *(bottom)*, whose later elegance is not evident in this photo.
OVERLEAF:
The most famous of an historic Italian line, the Isotta-Fraschini Tipo-8A had huge, straight, eight-cylinder engines of 7.4-litre displacement and 160 horsepower. Sadly, this luxury car ended production in 1932, and sadder still, unsold cars were *junked* in 1939.

THE RATIONAL PERIOD (1916–1926)

As the Primitive period drew to an end, the way to genuine mass production was pointed by Ford's climb from producing about 18,000 cars in 1909 to producing half a million in 1915. The Model T was leading the United States automobile industry, not by its styling, which was merely adequate (especially in retrospect), but by its utility and low price.

The Model T drew many famous names after it. The Willys-Overland firm soared from 4,860 cars made in 1909 to second place behind Ford with 92,000 in 1915. When the Dodge brothers grew tired of their relative anonymity in supplying engines to their sometime friend Henry Ford, they decided to build their own automobile and jumped into the market in 1915 with 45,000 cars. Their car was exactly what might have been expected from the hard-drinking, hell-raising Horace and John Dodge: a well-built vehicle of rugged all-steel construction, with rakish fenders that complemented the softened contours of the body. It was

"I'm telling you that needle's around here someplace!" In the Nebraska grasslands of 1916, two cars in search of a haystack—or a road.

characteristic of the cars that would help the industry leap from a total of 2 million vehicles in 1920 to over 4½ million in 1929.

One result, of course, was the wholesale disappearance from the marketplace of manufacturers who could not expand their production and marketing sufficiently to keep pace with the industry. The pressures were especially severe on companies that manufactured unusual cars or insisted on making all of their own components. As a result, the world lost such gems as the rotary-engine Adams-Farwell ("It Spins Like a Top"). This was no Wankel rotary, with rotors all neatly enclosed, but an engine similar to the aircraft Gnome-Rhone, in which the cylinders whirled merrily around a fixed crankshaft.

Gone, too, was the handsome and well-respected American Underslung that had all the benefits of a lowered center of gravity that would be claimed by later Cords and Hudsons. Crouching like a runner in the starting blocks, with its front straight-slanting fenders cradling huge drum headlights, it was extremely stylish in both its conventional and underslung-frame versions. American sold more than 45,000 units in its nine years of production before failing for financial reasons.

Also among the missing would be the peripatetic Cameron, which in its seventeen years of existence produced cars powered successively by steam, gasoline, air-cooled, and water-cooled engines. In the process the company moved like a man avoiding a subpoena, from Rhode Island to Massachusetts to Connecticut to Michigan to Ohio. Perhaps the raciest of the Cameron lot was the 1912 Yale Featherweight Flyer, with its pointed racer nose and natty rumble seat. They neither make them nor name them like that anymore.

Imagination didn't die in the next decade, although many more car manufacturers did. The innovative Briscoe, for example, lasted from 1914 until 1921. Where today could you buy a car with such advanced ideas as a single cyclopean headlight, a papier-mâché body, and the option of trying a four-cylinder engine that could be exchanged in thirty days for a V-8 if you so desired? Almost 40,000 Briscoes were built during the period; ironically, the company failed when it turned to production standardization in the hope of achieving a stable competitive position at its relatively low production rates.

Above all else, production standardization was responsible for lifting other manufacturers to new heights of output. The trend was made more difficult by the steady addition of equipment we now take for granted. The self-starter, introduced by Cadillac in 1912, had become universal. Roll-up windows and windshield wipers were the rule, although the wipers were often still manually operated. Four-wheel brakes were coming into the marketplace. Alternative power sources—electric or steam—were no longer a genuine option, and engines had grown fairly standardized into a few basic types. Suspension was equally uniform in the use of leaf springs of various configurations. Rearview mirrors became standard, and provisions were made for cigarette lighters and ashtrays.

Automobile accessories of a thousand types were available for individual owners to customize their cars. You could obtain everything from a raccoon-tail streamer to an elaborate water-level gauge that could be installed in the radiator cap. Styling was given an assist by the introduction of DuPont's synthetic Duco paint, which permitted shorter paint-drying times to match faster assembly schedules. The reason Henry Ford had insisted on producing only black Fords was not that he hated colors; it was that black was the only color

Marie Prevost, a Mack Sennett Bathing Beauty before she became a leading lady, is shown in the sporty Buick roadster she used to drive between the De Mille Studio and her beach cottage.

capable of drying in a few hours, accommodating Ford's frenetic production pace. When Duco came along, manufacturers had a wider choice, and even Henry finally offered a variety of colors.

Already, you could buy a car on a time-payment plan.

In view of the ongoing dual process of standardized production and expanded equipment, it was natural that styling would itself be standardized. Even as this happened, a revolution occurred at General Motors that would have far-reaching effects on the styling of all cars. The pell-mell changes in command that had seen William C. Durant seize and lose power—alienating major automotive figures like Charles Nash and Walter Chrysler as he did so—also saw the advent of a new manager, Alfred P. Sloan.

Sloan would rationalize the giant corporation into the most fearsome competitor in the automotive marketplace, and he would use styling changes as the instrument to force annual model changes upon an accepting public and a reluctant industry. At the same time, he created from the chaos Durant had left in his wake the image step-ladder of Chevrolet, Pontiac, Oldsmobile, Buick, and Cadillac. Sloan hoped that every man in a new job would buy a Chevrolet and then tell the world of his success, step by step, until he became president of his firm, driving a Cadillac. Sloan's planning had enormous impact around the automotive world for the next fifty years.

But in the Rational period, before Sloan's magic could take hold, the most notable instrument in the creation of the styling imprint of the time was the ruler.

We can compare three of the leading makes during the period to verify the dominance of the straightedge over the French curve. A useful base is the Model T, which changed in appearance only slightly as it sold by the millions from 1915 to 1925. In a 1915 sedan, the angular engine cover butted into a slanted coaming, which in turn attached to the square, high body. The 1925 edition was essentially the same, except that the hood was more rounded and the coaming was somewhat conformed to meld into it.

The Willys-Knight touring car serves as the next example. The typical Willys of the period did not resemble the comic little bug-nosed rascals of the pre–World War II era, nor the sleek Aero series styled by Phil Wright in the early 1950s, nor yet a Jeep. Willys was then a full-line manufacturer, using the highly respected and widely licensed Knight sleeve-valve engine. The Willys-Knight touring car of 1916 was handsome enough, with the top line of the doors suppressed slightly below the top of the hood line. But by 1921, you could draw a straight line from the radiator, across the hood, cowl, and doors, to the padding of the convertible top. Willys did make a concession to streamlining by slanting the windshield back slightly, but production considerations clearly dominated styling in general.

Similarly, the Buick's styling changed from the molded contours of the 1915 hood and bonnet, with a four-door body that tapered from the smooth cowl line outward to a curved rear, to the rigid, level, four-square styling of 1925.

Virtually forgotten now, Overland was a famous name in the automobile world from 1903 to 1926 (in 1925, 215,000 Overlands were produced), and it even had a brief resurgency as a Willys model in 1939.

OVERLEAF:
Perhaps one of the best things Lee Iacocca did was to inspire a host of Mustang imitators. Among the most successful, and certainly one of the most beautiful, was the 1967 Chevrolet Camaro. Although the base price at the time was only $2,466, it was possible to spend twice that by adding goodies the dealer tempted with.

The cumulative effect was that cars of the Rational period were even more difficult to distinguish by make than cars are today. The contemporary problem is the plethora of models and their overlapping price structures. In the Rational period, the confusion stemmed from straight sides, square fronts, square rears, and round cycle-type fenders on virtually every make.

In Europe, where production volumes remained at or below prewar rates in the United States, there was a great deal more variation, with some gorgeous (if occasionally idiosyncratic) results. Tiny cyclecars were very popular, and some, like the Morgan, grew to be great. The Morgan started out as a three-wheeler that looked more like an ice cream vendor's cart than a car, and the firm stayed with the three-wheel formula through 1952. But it was a fierce little rasper with great performance and cars with a conventional fourth wheel were also built, beginning in 1936. The Morgans most familiar to Americans were produced after World War II in the 4/4 series. The no-nonsense conception of the Morgan—with its seats mounted directly on the floor plan, its stiff springing, and its gutty engine—carried over to its styling, which was tougher and more pugnacious than the more familiar M.G. TF.

The Tamplin, which looked as slender as the Model T's that Laurel and Hardy used to compress between streetcars, derived its mechanical components directly from motorcycles. Its great virtue was that it sold for only £150 (about $750 then), but it was somehow also a handsome car by any standard other than size. In Italy, the Lancia Lamba complemented its radical unit-body construction and independent front suspension with torpedo-slim styling. The German Audi of the period was handsomely styled, deftly incorporating the V-shaped radiator that seems to have been ubiquitous on German cars of the time. It also achieved a less ponderously Teutonic look by carefully and harmoniously carrying through its belt and fender lines.

The Lagonda and the Aston Martin presaged

Overland
M-650

BLUE PLATE
SPECIALS
WAITRESS WANTED
PEOPLE SKILLS NOT REQUIRED
FOUNTAIN
Creations
SS
camaro
350

two decades of classic British automobiles. The Lagonda concealed its highly sophisticated engineering (hemispherical combustion chambers, twin overhead camshafts, and four-wheel brakes) beneath a luxurious body that sported a handsome V-shaped windshield. The Aston Martins were sleek, hell-for-leather cars equipped with exhausts channeled back in great pipes, tiny cycle fenders, and great wide belts that secured the bonnet to the chassis. For the next fifteen years, British stylists would look to these cars for inspiration.

For the most part, though, the English and European roads were filled with the same slab-sided, strait-laced, easy-to-produce, wheeled boxes that predominated in America. Fortunately, a smiling promise of the future was conferred upon the most expensive cars by the great *carrossiers*—carriage makers of proud heritage who had served royalty and the wealthiest of families for generations, but now were having to change with the times.

Many of these elite carriage makers elected to become full-scale manufacturers of complete automobiles. Others preferred to stay in a familiar field, building custom carriages to suit the specifications of the rich. Later, when styling trends so dictated, new firms unburdened by carriage-making tradition were established. Just as classic carriage body styles have been preserved in automotive nomenclature, so have the names of some of these carriage builders, as with the Cadillac Fleetwood and the Chrysler LeBaron. These names and others—Derham, Brewster, Brunn—were once as distinguished as those of the car makers themselves.

The great coach builders catered to the wealthy, but their influence benefited everyone. A custom-designed body that caught the eye of passersby usually caught the eye of company designers, too, and soon appeared in a variety of makes. In 1925, Lincoln began the practice of hiring a design firm to create bodies for their production automobiles. Until then, companies had relied entirely on their own resources, occasionally availing themselves of a consultant. And despite the efforts of individual geniuses, the constraints of the time, in terms of both tradition and manufacturing capability, resulted in many cars' looking alike. Examining the 1928 Hupp and Graham sedans would inevitably lead you to think that they came from the same person: bodies, fenders, wheel discs, moldings, bumpers, radiator shells, headlamps, and even the parking lights look interchangeable. Yet they were designed in complete secrecy and with total independence. The Hupp was the product of Amos Northup, using a Murray body, while the Graham-Paige had been designed by R. L. Stickney and Hugo Pfau of LeBaron.

A sporting king—who loved Hispano-Suizas—Alfonso XIII of Spain saw his country safely through wars and revolutions, but quietly decamped from Spain, without abdicating, in 1931.

One characteristic of the 1916–1927 time period was the prestige manufacturer whose low volume of production resulted in sales that were confined to a relatively small area. Such car makers could not afford—and in fact, disdained—to use a large dealer network, preferring instead to concentrate on building the best car they could in small numbers. John B. McFarlan (who also created one of the world's first industrial parks) built the luxurious McFarlan in small numbers for almost eighteen years. The best year (1922) saw 235 cars leave McFarlan's factory, each one constructed with painstaking care. While the early cars clearly showed their carriage factory heritage, sleek departures soon became evident, where sheet metal was molded to merge with large glass areas. The McFarlan combined a heavy Rolls-Royce elegance in its body with a delicate simplicity in its fenders (usually rendered as simple arcs). But it was the overall presence of the McFarlan that was so impressive: when it hove on the horizon, you knew someone wealthy and important was approaching.

A galaxy of automotive leaders created the ReVere in 1918, a car with both Duesenberg and Stutz heritage. Low and fast enough to be purchased by no less an automotive connoisseur than King Alfonso XIII of Spain, the ReVere was a superb car wrecked by an unfortunate management problem that smacked of fraud. It was a shame, for the ReVere's ground-hugging style was, as the firm's advertising repeatedly asserted, "classy."

The best representative of the type, however,

was probably the Cunningham, which was built in small numbers in Rochester, New York, between 1907 and 1936. The Cunningham was like the McFarlan in its burly, self-assured styling: a Harold Lloyd turned into a Tom Mix driving a Cunningham. A wide variety of body styles were available—as might be expected from a firm that previously built not only carriages but dog carts, boats, and sleighs. Perhaps the handsomest of all were the long-hooded, boat-tailed speedsters, in which the spare wheels were mounted rather far aft, adjacent to the cockpit. While the clientele consisted largely of wealthy folks from the eastern seaboard, the Cunningham V-8 was also purchased by many famous Hollywood personalities, running the gamut from petite Mary Pickford through discreet Cecil B. De Mille to hardly effete Fatty Arbuckle.

In Europe, a similarly benign climate that fostered quality in unique formats prevailed for a considerably longer time. France's essential Gallic spirit, Gabriel Voisin, produced a long series of aerodynamic (if sometimes eccentric) cars that seemed to match his own irascible and volatile personality. He didn't like fenders that vibrated, so he attached distinctive aircraftlike struts that ran like a strap over the V-shaped radiators. Similarly, he considered that cars were primarily for driving, not riding in, so the passenger compartments were squeezed between hoods that went on forever and massive trunks. Additional luggage space was offered by huge panniers, mounted left and right. Yet whatever individual peculiarities certain individual features might possess, in the last analysis the Voisins were good-looking, racy cars.

The Voisins were complemented by the work of the Sizaire brothers, who tried virtually everything to win over moneyed customers (including making a replica Rolls-Royce, which they advertised to have better performance for less money). At the same time, smaller quantities of luxury cars were produced by more famous names such as Hotchkiss, Lorraine-Dietrich, Darracq, and Bugatti—each of them utterly distinctive. The Hotchkiss, more familiar as the name of a machine gun, was essentially conservative, offering ample room and a special glass divider (a windshield, by any other name) for the rear-seat passengers. The Lorraine-Dietrich was a sports car of Le Mans–winning quality, its bodywork imparting a supple impression that was unusual for the time. The Darracq A type followed American luxury styling somewhat, but embellished it with a towering body and a latticed, V-shaped windshield.

And the Bugatti? Always distinctive, from tiny elemental racers to huge extravaganzas like the later Royale. But the overwhelming sensation imparted by a Bugatti was its utter integrity and cohesion—its styling almost always forming a taut minimalist skin over the intricate engineering.

In smaller countries, protective tariffs and national pride supported native luxury cars for many years. Belgium's Minerva and Imperia makes were world-famous, and many aficionados ranked them with Isotta-Fraschini and Mercedes, if not with Rolls-Royce and Hispano-Suiza. Their distinction was conferred not so much by the flowing lines as by the impressive size and the intricate detailing, for every element in both cars was superbly executed.

The generally poor roads of Central Europe challenged manufacturers of luxury vehicles to the limits of their engineering skill. In Austria, both the Austro-Daimler and the Graf & Stift featured independent rear suspension. The Austro-Daimler used the *de rigueur* V-shaped radiator of the Germanic countries to cool its very modern aluminum monobloc engine (designed by Ferdinand Porsche). Its lines were essentially straight, but the accents of the rakish fenders and the luster of the marvelous paintwork underscored the fact that the Austro-Daimler was clearly an ultra-luxury car.

The Graf & Stift was also of extremely high quality and had an added mystique from the fact that it was in a 1914 model that the Archduke Francis Ferdinand was assassinated in Sarajevo. The archduke's car may still be seen in the Vienna Museum of War History. The first make of car to

be produced in Austria, it was equipped with front-wheel drive by 1897; still, the Graf & Stift's general lines were similar to those of American luxury cars of the period.

The Czech Tatra was nearly always radical in appearance and in engineering, offering a backbone tubular frame and independent suspension as early as 1923. The lines were spare, and the

LEFT:
This beautiful Packard Twin Six sedan *(top)* was one of 5,193 produced in 1920. The durable V-12 engine produced 90 horsepower. In 1921 it was in just such a car that Warren G. Harding rode to his inauguration, becoming the first American president to use an automobile for this event. The distinctive Packard radiator styling first appeared in 1904. The 28-horsepower, 4-cylinder engine powered everything from a runabout to a limousine, with this 1904 touring car *(bottom)* falling somewhere in between.

RIGHT:
If there had been a *Consumer Reports* at the time, the 1919 Cunningham *(top)* would have gotten high marks for reliability, quiet operation, and safety. Add the fact that it was completely handbuilt, however, and it's no surprise that Cadillacs could be had for half that price.

Early in this century there were many Pope cars—the Pope Robinson, the Pope-Toledo (the Mile-A-Minute Car), Pope-Tribune, Pope-Waverly—but the most famous was the Pope-Hartford *(bottom)*, manufactured from 1904 to 1914. It was one of the very few cars for which all components were manufactured in the parent factory.

fenders were mere hints of the stylist's pencil as they curved back to join in a short running board. Later, as we'll see, the Tatra led the way in the development of truly streamlined bodies. The Skoda provided a conventional counterpoint to the Tatra, offering conservative models across the board but maintaining a high level of quality that reflected the firm's early association with Hispano-

Suiza. Tatra and Skoda are still well-known names for Czech cars, but others (including the Laurin & Klement) were also internationally renowned.

England, of course, remained preeminent in the world of traditionally styled cars of very high quality and (sometimes) eccentric engineering. The Rolls-Royce, as always, set the standard for the most upright styling, as it did for the most superb engineering; however, the Leylands, Daimlers, and Bentleys followed suit at no great distance. The manufacturers provided good solid chassis and engines, and if the customers didn't care for the conventional bodies (which tended to be tall, bulged, and encumbered with external gas tanks and boxes built into the running board), they could go to the great custom firms and choose their own special bodies.

Hans Ledwinka designed cars as early as 1897, but will always be remembered for his distinguished Tatras, manufactured in Czechoslovakia. This 1937 model featured an air-cooled V-8, placed in the rear. It was a fast car, but somewhat dangerous: at one point, German officers were forbidden to drive Tatras due to their treacherous handling.

The Baroque Period (1927–1942)

Just as aviation historians dispute which period was its golden age, so do automotive historians advocate different decades as being the golden age of cars. Most would agree, however, that some of the most beautiful custom and production cars in history were produced in the fifteen years following 1927.

At the time, the fabric of the automobile world was tearing. On the one hand were manufacturers that were doomed to go out of existence as a result of inadequate production, styling failures, or management fatigue. On the other hand were a smaller number that would manage to become industry leaders in both production quantity and (to a lesser degree) engineering and styling. General Motors and Ford showed an increasing reluctance to get too far in front; one safe gray-flannel motto of the time was "Never first, but always a better second."

Chrysler took a different and almost fatal tack by matching its engineering prowess with the late and generally unlamented Airflow line. Walter Chrysler loved cars and knew how to run companies; best of all he knew how to take advice, and the combination of qualities permitted him in a single year to transform a moribund Maxwell/Chalmers combine into the sensational Chrysler Corporation. His first car in 1924 incorporated a bonanza of engineering achievements, including full-pressure lubrication of the high-compression engine and four-wheel hydraulic brakes. There followed ten years of uninterrupted successes that saw him acquire the venerable and prestigious Dodge Company and introduce the Plymouth and De Soto lines.

Chrysler cars were engineering and styling masterpieces—the luxurious Imperial using Locke, LeBaron, and Dietrich bodies and influencing designers all around the world. But the problem with great performances is always what to do for the encore, and in 1934 Chrysler's encore was the Airflow.

THE AIRFLOW EXPERIENCE AND GM HEGEMONY

Walter Chrysler had benefited from the advice and counsel of his "Three Musketeers": calm Owen Skelton, roly-poly and devout Fred Zeder, and passionate car nut and intellectual Carl Breer. They had joined ranks with Chrysler as engineers in 1915, and they worked together in an almost mystical communion of their complementary engineering minds.

Zeder served as the nominal leader of the group, always prodding it to advance farther along the cutting edge of automotive technology, while remaining utterly dedicated to quality, safety, and reliability. In contrast to virtually every other Chrysler executive (and for that matter, to every other automotive executive), Skelton was a fashion plate in dress and had a presence of almost movie-star quality. He experimented as early as 1928 with rear-engine cars and was responsible for developing four-wheel hydraulic brakes and the famous Floating Power rubber engine mountings. Breer was small, dapper, and intrigued with everything automotive. While his technological insight was as keen as Zeder's and Skelton's, his real gift was in making intelligent practical application of the engineering innovations to production.

The three had given Chrysler its engineering advantage, and they would also create the Airflow. Developed over a six-year period, the car was cursed with the same fault that plagues modern aeronautical design: hobbyshopitis. Determined to build a low-drag, aerodynamically efficient vehicle, the three men also wanted to make advances in suspension, weight distribution, and power. Like the apocryphal medical case, the operation was a complete success, but the patient died.

The cause of death was ugliness. When the Chrysler and De Soto Airflows were unveiled in 1934, they did ride better than other cars, and they were both powerful and quick. Drag was substantially reduced, and the automatic overdrive produced sensational mileage. Drivers and passengers had lots of room and were cradled safely and comfortably between the wheels in a semiunit body and frame. The half-walnut design of the exterior shell, however, was rejected by the public, and for a very good reason: the Airflow was unattractive and not susceptible to improvement. Fortunately for Chrysler, the error had not been forced on the Plymouth and Dodge divisions, which sold 392,000 units while De Soto and Chrysler jointly sold about 39,500.

Pride and investment dictated that the line be continued for three more years at Chrysler and two more at De Soto, but after 1934 the manufacturer's emphasis was shifted to cars of conventional styling, which were termed Airstreams in a futile attempt to save face.

The Airflow experience perhaps explains why,

during the first part of the Baroque period, smaller firms tended to lead with styling changes, which General Motors and others would then follow. This trend is exemplified by the work of the great Amos Northup. In 1931 he had revitalized Reo with the famous Royale, which turned the heads of the relatively few people who saw one of the several thousand that were produced. Northup's tour de force is all the more remarkable because it was neither a radical departure from the style of the times nor drastically more expensive. Instead, he softened the body surfaces generally and subtly, while imparting a rakish line to the fenders. The accents were striking: a V-shaped radiator grille; long, streamlined headlights matched by parking lamps on the fenders; and a simplified windshield. Above all, the styling was clean and crisp, with all elements crafted in smooth harmony.

Working primarily for Walter Murphy's custom body firm, Northup was graciously permitted to free-lance. The following year, he followed the Royale with the Graham Blue Streak. The Blue Streak's rakish fenders were skirted, the lower elements being drawn down so that they formed a colorful outer shield where formerly the soiled linen of oily chassis had been displayed. The broad, gleaming grille was slanted back, giving the car a sense of motion even at the curb. It also covered elements of the chassis and imparted a unified look to the front of the car.

By the mid-1930s, the mantle of styling leadership had passed to the General Motors Art and Colour Department under Harley J. Earl, which began to dominate the industry in a way never before considered possible. This was due in equal parts to the genuine quality of the styling, to the sheer quantity of GM cars on the road, and to the volume of GM advertising. The public became mesmerized with the idea that the GM shape—whatever it was—was the shape that cars ought to be. A second pattern of leadership emerged, in which the General Motors design elements were copied to a great degree by Ford and Plymouth, and as much as was economically possible by the independents. This domination accelerated during the Escapist period, but by 1937—the middle of the Baroque period—it was already well-established. What was true of style was true to a lesser degree of mechanical elements, which were easier for the smaller companies to finesse. A car maker could raise the compression slightly in a fifteen-year-old engine design, call it a "New Red Flame Eight," and stand a fair chance of getting the public to buy it. But there was no way to advertise your way out of an old-fashioned body style, although efforts to do so were unstintingly made.

Detail of a 1921 Lincoln Phaeton Model L, with coachwork by Brunn. Henry Leland was a fantastic figure, one who had the drive to leave Cadillac at age 74 to start the Lincoln Motor Company where he built Liberty engines before launching a prestige automobile.

Harley Earl essentially dictated that all cars should look as much like the current Cadillac as their price range permitted. There was a sequenced phasing of the Cadillac's style elements within the GM lines. Senior Buick and Oldsmobile models could be expected to share a new belt or fender line, for example, but wouldn't be allowed to emulate the grille shape until the following year. Similarly, reflections of the more expensive cars would appear in the lower-end Pontiac and Chevrolet lines—usually in the parts for which a larger manufacturing scale resulted in economical production.

The next year, more of last year's Cadillac features would trickle down toward the Chevrolet, even as new elements were added to the new year's Cadillacs. It was a comforting, predictable arrangement, and GM products naturally had the inside track in following it. The other manufacturers would use as much industrial espionage as was available at the time to gain insight into the GM styling, but their catch-up process was always more expensive because it had to be done on a crash basis.

Harley Earl started the process subtly with the 1927 LaSalle, which artfully drew on the Hispano-Suiza for its lines. It was so successful that his services were in demand by all of General Motors, and the Art and Colour Department was formed in 1930 to dictate styling for the entire lineup of cars. Under this system, production costs could be stabilized even while changes were taking place, because of the sheer predictability of the changes. If Cadillac emerged with fenders that flowed through the doors in 1941, it was inevitable that

Attention to detail made the Duesenberg the frequent choice of royalty the world over. For example, when Maharajadhiraj Raj Rajeshwar Sawai Shree Yashwant Rad Holkar Bahadur of India commissioned the prestigious coachbuilding firm of Gurney Nutting to body the last Duesenberg chassis to roll off the assembly line (1936), he wanted some special touches, including the installation of red and blue lights in the front fenders *(top two)*: The red light meant the Maharajah himself was driving; the blue was used when the Maharanee was behind the wheel.

***(Third from the top)*: The chrome trim on the wheel of a 1936 Duesenberg convertible wheel. Wire wheels were beautiful, but difficult to maintain and a devil to restore.**

No one—not Fred, not even Augie Duesenberg—ever predicted that the winged Duesenberg logo *(bottom)* would mean—almost automatically—a million dollars or more in automotive worth.

OPPOSITE:
Gordon Buehrig, famed for his design of the 1936 Cord, did excellent work for Duesenberg as well, as this 1933 Weyman-bodied SJ Speedster reveals. The boat-tail motif, as incongruous as it might be on an automobile, never failed to please.

lesser GM cars would follow suit in the near future. There was a progression even of detail: headlamps moved like the migrating eyes of a flounder, from being stuck on chrome bars in front of the grille to being nestled on the side of the hood to being slumped down in the sinuous curve between hood and fender to being locked flush into the fender itself.

It is a little disheartening to examine the 1937 Hudson, Nash, Studebaker, Plymouth, Dodge, De Soto, and Chrysler and see how closely they tried to ape the GM formula. Even sadder is the fact that none succeeded. Imitation may be the sincerest form of flattery, but it is the most difficult way to excel. Each of the imitations bore a false note: the Chrysler products were a shade too rotund, rounded where they should have been crisp; the Hudson looked bulbous and fussy; the Nash was generally a better effort but was spoiled by a poor grille and a squint-eyed set of headlights.

The one success was Studebaker, which not only created the best of this subgroup of 1937s, but broke away in subsequent years to chart—with Raymond Loewy's help—a distinctively beautiful path of its own. Loewy was not encumbered with the industry's tradition of car design, nor was he intimidated by Studebaker's accountants and engineers. The Vice President of Sales, Paul G. Hoffmann, respected Loewy's artistic ability and gave him a free hand to impart an elegant simplicity to the pontoon fenders and rounded grilles. Clearly, Loewy knew how to recognize situations where less was more; and by 1939, his styling efforts had brought Studebaker back from the brink of bankruptcy to eighth place in sales, with 106,470 automobiles sold.

ACTIVITY OUTSIDE THE GM ORBIT

Until well after World War II, Ford had the strength of will to persist with its own design formulas, while Packard remained above the fray. There is an old aviation aphorism that an aircraft that looks right flies right. The Packard perfectly vindicated the automotive version of the phrase. Packards, with

Omaha, Nebraska—November, 1938. It was the month that Kate Smith first sang "God Bless America," Pearl Buck won a Nobel Prize, *Gone With the Wind* was stunning moviegoers—and cars looked as much alike as they do today. *(Photograph by John Vachon)*

their high-cheek-boned grille, had a slim, crisp, uplifted, Katharine Hepburn-type beauty that always looked right. Now remembered chiefly for this beauty, they also had superb engines and handling qualities far superior to contemporaneous Pierce-Arrows and Cadillacs.

After 1939, the future was given a trial in dream cars, those no-costs-barred, no-engineering-barred fantasies that grew over time from merely extravagant cars that could still be driven to incredible shapes presaging the future—not as it could be, but as sci-fi artists would like to imagine it. The first in this long line was the Buick Y-job that presaged Buick styling until 1958. Other major companies created similar dream cars, although few enjoyed the same consistent success that General Motors did.

But despite GM's dominance, other stylists presented designs that were appreciated, even if their influence was not as widespread as Harley Earl's. The first of these notable designs was by Phil Wright, who later designed the postwar Willys Aero series. Wright scored a smashing success with the aerodynamic 1933 Pierce Silver Arrow, which the press agents boosted as being "born in a wind tunnel, and made by hand." The car, the sensation of the 1933 Chicago World's Fair, combined tradition and modernity. The unmistakable Pierce-Arrow grille and fender-mounted headlights were beautifully integrated into an ultrastreamlined envelope. The all-steel turret-topped body was slab-sided, with recessed door handles and enclosed running boards. The doors themselves were so thick that they cut down on interior space, but they melded neatly with the front fenders. The rear wheels were enclosed by a skirt, and the spares—instead of being located in the traditional side-mountings—were enclosed within the front fender. Both the windshield and the rear window were V-shaped, and the body tapered to create a sinuous V between the pontoon rear fenders, which bore the traditional Pierce-Arrow three-lamp taillights.

Only five of these cars were built, and relatively little of the styling survived in the production Pierce-Arrows that followed (in ever-diminishing numbers) until the last virtually hand-built car was rolled out in 1938.

In a less ambitious undertaking, Raymond Loewy, the old damage-control and rescue specialist, created a car aesthetically midway between the Chrysler Airflow and the Pierce Silver Arrow, in the 1934 Aerodynamic Hupmobile. He retained the Amos Northup look, but with pontoon fenders, to make the Hupp easier on conventional eyes. The windscreen was a unique three-piece unit lifted from the German Horch but somehow harmonious with the whole. Unfortunately, the new styling didn't matter, for Hupp production had dropped to less than 10,000, and the firm could not be saved from extinction.

There is no little irony in the fact that the man who did the best with the aerodynamic look was the namesake for the greatest postwar failure: Edsel Ford. Deeply conscious in 1932 that the great Lincoln motorcar was heading for oblivion just as Marmon, Peerless, and so many others had, Edsel Ford was attracted to an advanced concept put forward by a young Dutchman named John Tjaarda. Tjaarda had worked with the Fokker Aircraft Company in the Netherlands and later with the famous aerodynamicist Dr. Alexander Klemin of New York University. (Klemin had also worked with Chrysler, and some critics blame him for miscalculating the strength of the Airflow chassis and requiring it to be far heavier than was necessary.) Tjaarda was employed by the Briggs Manufacturing Company, which supplied bodies to both Ford and Chrysler. The designer had long thought about "the ideal car," which he thought would be streamlined, have a unitized body and frame, and be inexpensive to produce. He showed sketches to Edsel Ford, who authorized him to proceed in total secrecy from the hidebound senior management of both Ford and Briggs.

The result was the Lincoln Zephyr, a fully streamlined car that used aviation principles in its semi-unit construction as well as in its design. The public liked the Zephyr much better than it had the Airflow series, and sales figures shot to ten times the 1932 Lincoln totals, with 17,715 sold in 1936 and then a super-record 25,243 sold in 1937.

Styling was further improved in 1938, and the influence of the Zephyr began to be visible in both the Ford and Mercury lines; nuances of styling similarity also appeared in the Studebaker. Auto iconoclasts—car blasphemers all—contend that Ferdinand Porsche's imagination was affected by seeing Tjaarda's early work, and that this influence carried forward into the Volkswagen. The assertion gains a bit more credibility if you consider that Tjaarda had proposed a rear-engine, unit-body, beetle-shaped design in one of his Zephyr studies.

Normally a smash styling success like the Zephyr is a one-of-a-kind thing: a height is reached, and then a gradual deterioration occurs as changes are made over the ensuing model years. But not only did the Zephyr improve in appearance from year to year, it spawned what many people believe is—with the Cord—one of the two most beautiful American production cars ever, the Lincoln Continental.

The Continental was once again fostered by Edsel Ford, adding his own touches to the beautiful lines generated by Bob Gregorie, who had been instrumental in refining the Zephyr. Although the Continental was originally intended to be Edsel's personal car, the public response to it was so favorable that large-scale production of the car

OVERLEAF:
This snippet of Duesenberg door *(left)* belongs to the short-wheelbase version formerly owned by Gary Cooper.

The leather-trimmed door panel of the famous, colored-light-coded 1936 Duesenberg SJ owned by an Indian Maharajah. He owned thirty of the world's greatest motorcars, and housed them on his 10,000-acre estate—on which there was no suitable road. The answer was simple: build a forty-mile-long road of crushed white rock for them.

began in 1939. For reasons of economy, much standard Zephyr tooling was used, but so artful were the cuts and insertions that the Continental had an unforgettable look of its own.

Unlike that of the Chrysler, the engineering of the Zephyr and of the Continental was rather mundane, using transverse-spring suspension and bolting on four more cylinders to convert the Ford V-8 into a V-12. But the cars were light and performed well. And best of all, they gave the kiss of life to a make well worth saving.

THE BEAUTIFUL FEW

Other handsome American cars of the period were made in much smaller numbers (in some cases, only one or two of a kind). The Marmon, which had been an excellent automobile for thirty-one years, finished out its lifespan with brilliant V-16 production cars and a final, gallant V-12—a thoroughly modern car with a remarkable suspension system. As the Stutz became more and more handsome, its production declined to single-digit quantities. In 1934, Stutz offered no less than thirty-five body styles in six series of cars, and only six cars left the factory! Neither nostalgic names like Bearcat, Speedster, and Torpedo, nor DV-32 engines—four valves per cylinder—could sustain it.

LEFT:
One of the great ironies of automobiledom is that Edsel Ford, the man directly responsible for the beautiful Lincoln Zephyr and Continental, will be remembered most for the postwar car that bore his name—one that he surely would have aborted.
OPPOSITE:
The distinctive fender headlights of the Pierce Arrow were introduced in 1913, some thirty years ahead of most makes. Shown here is the 1931 model, but the Pierce Arrow was always a car of superb quality. *(Photograph by Margaret Bourke-White)*

But in the hands of an aggressive businessman and supersalesman like Erret Lobban Cord, styling could lift sales to new heights and automobile bodies to an art form. Cord had a justifiably unlimited confidence in himself. As the world's best salesman of the Moon automobile (Moon was an assembled car of good quality built in small numbers in St. Louis), Cord was approached by some concerned Chicago businessmen to see what he could do for the Auburn Company, which in 1924 was teetering on the brink of oblivion. Among the firm's ''assets'' were 700 unsold and totally undistinguished cars.

Cord accepted the job without a salary, but with the option to purchase control of the firm if he saved it. He immediately improved the cash-flow situation drastically by slapping dramatic coats of paint and a little additional nickel plating on the stockpiled Auburns and selling them. Next he added power, a commodity that the American public ardently desired but that the automobile companies seemed strangely reluctant to provide. In the process of installing bigger engines, he acquired Lycoming (like Continental, a supplier of engines to a number of assembled cars).

Auburn's sales jumped from about 2,500 in 1923 to 20,000 in 1929, and Cord's selling methods were so successful that he had to seek additional capacity, picking up factories the way Frank Lorenzo picks up airlines today.

In this regard, too, Cord had an excellent eye for what the public wanted. In 1926 he established a partnership with Fred and August Duesenberg, freeing them from the financial drudgery that had inhibited them in the past and giving them a charter to do just what they wanted to do: create the finest automobile in the world. Cord didn't believe that the Duesenbergs would make a great deal of money for him, but he knew that the association would flatter Auburn, which would make a great deal.

His success was far from being all razzle-dazzle. He knew he had to make attractive cars, but he saw to it that they were advanced in engineering

Hotel
BEVERLY
WILSHIRE

terms, as well. Alan H. Leamy, only 26 years old and a veteran of just eighteen months' experience in the Marmon styling department, presented himself to Cord because he heard a front-wheel-drive car was in the works. Cord liked his approach and his drawings, so he hired him on the spot. Leamy fashioned the first of the exquisite Auburn boat-tailed speedsters in 1928, and Wade Morton set a whole series of meaningful speed and endurance records in it. Cord saw to it that the speedsters had adequate power, Leamy made them look as if they were doing 100 mph standing still, and Morton drove them at better than 100 mph on the Bonneville salt flats. It was an intoxicating combination. In hard dollar terms, Cord had provided the Auburn with performance equal to that of the Bearcat for around $2,000—only 40 percent of the Stutz's price.

Cord went on building, and Leamy finally received his opportunity to work on a front-wheel-drive car, creating in 1929 the magnificent L-29. This classic masterpiece derived its low lines and long hood from having a front-wheel drive placed end to end with a Lycoming straight eight. It looked sensational, but it was expensive ($3,095) for a sedan and could achieve a maximum speed of only about 75 mph. Less than reliable despite a dedicated engineering effort and the consulting efforts of Harry Miller (the famous designer of front-wheel-drive race cars), the L-29 still sold over 5,000 units over a three-year period. E. L. had by this time acquired Stinson Aircraft and fostered there the building of the lovely Stinson Tri-motor, which he had used on his Century Airlines. Yet despite all this activity (much of which was concentrated on the manipulation of stock values), he still had time to grace the world with the marvelous line of Duesenbergs.

Only about 480 Duesenbergs were made in the eight years during which they were nominally in production, but their influence was felt around the world and persists to this day. The Duesenberg was certainly the only American car ever to be considered the full equal of, if not superior to, the Rolls-Royce. It was made without regard to cost: the chassis prices began at $8,500, later rose to $9,500, and finally reached $11,500. Several thousand more dollars could be spent on a "house" body designed by Gordon Buehrig, or as much as $25,000 could be laid out in selecting one of the premier coachbuilders and indulging in a little fantasy.

The real joy to be had from buying one of these cars was not the pleasure of flaunting one's wealth, but rather the satisfaction of expressing an inner sensitivity, an artistic flair. At least so the buyers felt as they were cozened in the Duesenberg plant—their every word agreed to, their every wish fulfilled. The agreeable servility of the sales personnel was in sharp contrast to the severity with which maintenance and driving instructions were imparted to both chauffeur and the interested owner.

The shattering beauty of the Duesenberg was complemented by its shattering records. These were at once capable road cars and very fast racers, hitting 100 mph in second gear. Top speeds varied with weight and tune, but 140 mph for stock vehicles was achievable. In 1935, Ab Jenkins drove a special-bodied SJ at more than 134 mph for twenty-four hours, averaging better than 152 mph during one hour, and edging toward 160 mph near the end of the day-long run.

Modern drivers testing restored Duesenbergs are invariably surprised at their relative docility in traffic, their good handling (without power steering, of course), and their fine braking. The key to the car's outstanding performance was the beau-

OPPOSITE:
A lovely 1930 Cord L-29 in front of the appropriately elegant Beverly Wilshire hotel. The L-29's radical front-wheel drive permitted a low silhouette that said "speed," even if it was not in fact a fast car. Five thousand and ten were built, and today one in good condition will command prices in the six-figure range.
BELOW:
A wheel detail from an L-29.

September, 1927. The streetsinger Luigi, at the Italian border. ***(Photograph by J. H. Lartigue)***

tiful straight-eight engine, descended through the years of racetrack refinement from a massive (and unsuccessful) Bugatti V-16 aircraft engine of World War I. Duesenberg engines featured double overhead camshafts and four valves per cylinder; extraordinary attention was given to balance and lubrication. At 4,250 rpm, the J produced 265 horsepower at a time when 100 was considered adequate for a luxury car. Hardly content, August Duesenberg designed a centrifugal-type supercharger that raised the power to 320. Besides allowing even higher speeds, the superchargers justified adding on four huge chromium-plated exhaust pipes that emerged from the hood and at night glowed red in the dark! The effect this display might have had on a young starlet being driven home by Cary Grant or Gary Cooper (both devoted Duesy owners) is simply too overwhelming to bear extended scrutiny. It's a wonder that the Hays committee didn't picket the Duesenberg plant.

Everything about the Duesenberg—frame, suspension, brakes, exhausts, instrumentation—was as well built as its engine. The Duesenberg was always thought to be a car for kings and movie stars, but we now know this isn't the whole story. The Duesy was a car for investors, too. Anyone who had shelled out the (perhaps) $20,000 necessary to buy one and then simply put it in a garage would have a $1 million treasure today.

You might think that a man who had revived Auburn, created the great line of boat-tailed speedsters, called forth the L-29, and permitted the Duesenberg brothers to do what no one else had done before might be content. Yet even as his empire was collapsing in the face of a relentless depression and antagonistic business rivals, E. L. Cord had one more beneficence for the automotive public: another front-wheel-drive car, again called a *Cord.*

Cord's financial situation was desperate: about the only money coming in was being drawn from such strange elements of his empire as the kitchen cabinet division. Yet he charged his engineers to

build upon the lessons learned with the L-29, and he called upon Gordon Buehrig to design an eye-stopping, head-turning body for it—on the cheap, for there was very little time and even less money.

It was another of those occasions on which economic stress forced designers to make the right decision. The famous classic hubcaps, much praised for their elegance, resulted from a need to maintain sheer low cost. Faced with the need to cover the wheels, Buehrig designed a simple outer ring, into which holes were drilled, surmounted by a perfectly plain hubcap. It was inexpensive to build but looked very expensive—exactly the two qualities Buehrig sought to capture.

The rest of the car was constrained in a similar way, and yet so great was Buehrig's talent that what emerged from his hasty process of drawing and mock-up proved to be a car that would win prizes all over the world and ultimately become the popular favorite classic of all time. In desperate haste, eleven virtually hand-built and completely unstandardized prototypes were completed in time to set the 1936 auto shows on fire.

This handsome 1934 Delage manages suggestions of both its Grand Prix-winning past and its prizewinning coachwork of the future.

It is difficult to say exactly why the Cord is so unbelievably good-looking, but it is one of those instances where each individual element—hood, front-wheel-drive housing, wheel covers, retractable headlights, taillight, everything—is marvelous in itself and even better in combination. Buehrig provided harmonious beauty, neither under- nor overstated, that was gorgeous from any angle, including (oddly enough) from directly below.

The coffin-nose Cord became an instantaneous public success. Despite its relatively high price of $3,000, orders poured in—orders that the company was unable to fill because there had not been time to work the bugs out. The cars that had been in the auto shows were not only undrivable, they were unusable, and all but two were cut up and junked. Over time, the front-wheel-drive speedsters with their Lycoming V-8 engines began to reach the customers; and then there was more trouble, for any new car has problems, and the independently sprung front-wheel drive had perhaps more than its share.

Production continued for two years, during which time the car's quality was steadily improved. But despite the introduction of supercharged models, Cord's luck, time, and money all ran out simultaneously. Only 2,320 had been built.

It was the end of the road for the Auburn-Cord-Duesenberg era—or so everyone thought. Prices on these cars plunged as they aged and parts became more difficult to get. I can remember a 1937 supercharged Cord Beverly sedan sitting in a filling station lot in 1944, perfect except for a cracked right front windscreen, with a $300 price tag and no takers. After the war, the historic, aesthetic, and nostalgic importance of the classic car was recognized, and prices began to climb. A pristine Auburn boat-tailed speedster or Cord now sells for as much as $150,000 and is a good investment at the price.

The Cord influence was maintained after 1937. Norman de Vaux—a sort of Joe Blftspk of the automobile world, who had failed successively with the Fageol, De Vaux, Continental, De Vaux again, and De-Vo—had purchased the Cord body dies. He worked out a complicated deal with Hupp and Graham, which added their debts together to produce in 1939 and 1940 a few Graham Hollywoods and Hupp Skylarks. Perhaps a total of 3,000 of these attractive cars were built, only to be generally ignored by the public. Today they are desirable as collectors' cars—their lower prices (an excellent example brings perhaps $11,000) reflecting the real degree of difference between a Cord and a copy.

Replica Cords were also made in the 1960s, with fiberglass bodies and assorted running gear. Somehow they just weren't what E. L. Cord had intended.

The die had been cast in America by 1934. After that date, production cars grew larger, sometimes handsomer, and always GMer. Perhaps the one breakthrough design of the period was the 1938 Cadillac Sixty Special, designed by William Mitchell, who ultimately succeeded Harley Earl. It was a fully integrated design, having the same essential rightness as the Cord but clearly pointing to the future that awaited all GM designs. The running boards were gone, and the body was made wider to accommodate six passengers. The body showed a minimum of chrome and a maximum of glass area, even though the number of windows on each side had been reduced from three to two.

The Baroque period was graced with other good-looking American cars. Packard, for example, retained its dignity and occasionally, with a Darrin special body, achieved greatness. But for the most part, the days of creative independence were already over.

Greetings from three stylish automobiles—the hood ornaments and grilles of a Packard five-passenger touring sedan *(top)*, a Pontiac *(middle)*, and a Lincoln *(bottom)*.

THE EUROPEAN BAROQUE

In Europe, the intense vitality and independence of designers produced a bevy of superb classics. There was a fundamental difference between the attitude of European owner/drivers toward cars and that of American owner/drivers. In the United States, cars had to be fast, beautiful, and dependable. Owners preferred to have nothing more to do with maintenance than to add fuel and oil with service periods at mileage intervals in the thousands. In Europe, a more intimate relationship existed between car and driver. The manufacturer considered that the driver (owner or chauffeur—it didn't matter) ought to expect cold weather to cause hard starting, and to anticipate lubricating valve trains thoroughly before starting, and to do a thousand other things that would have consigned the design to an American junk pile.

In return for extracting this measure of extra service, however, the car makers were able to provide more exciting machinery. The eccentricities of great designers such as Ettore Bugatti and W. O. Bentley were valued just as the eccentricities of Frank Lloyd Wright or Auguste Rodin were valued.

In our assessment of the European scene, a famous classic will be used to highlight the survey, while the general styling scheme will be examined in terms of more typical automobiles.

In France, the marvelously satisfying Bugatti

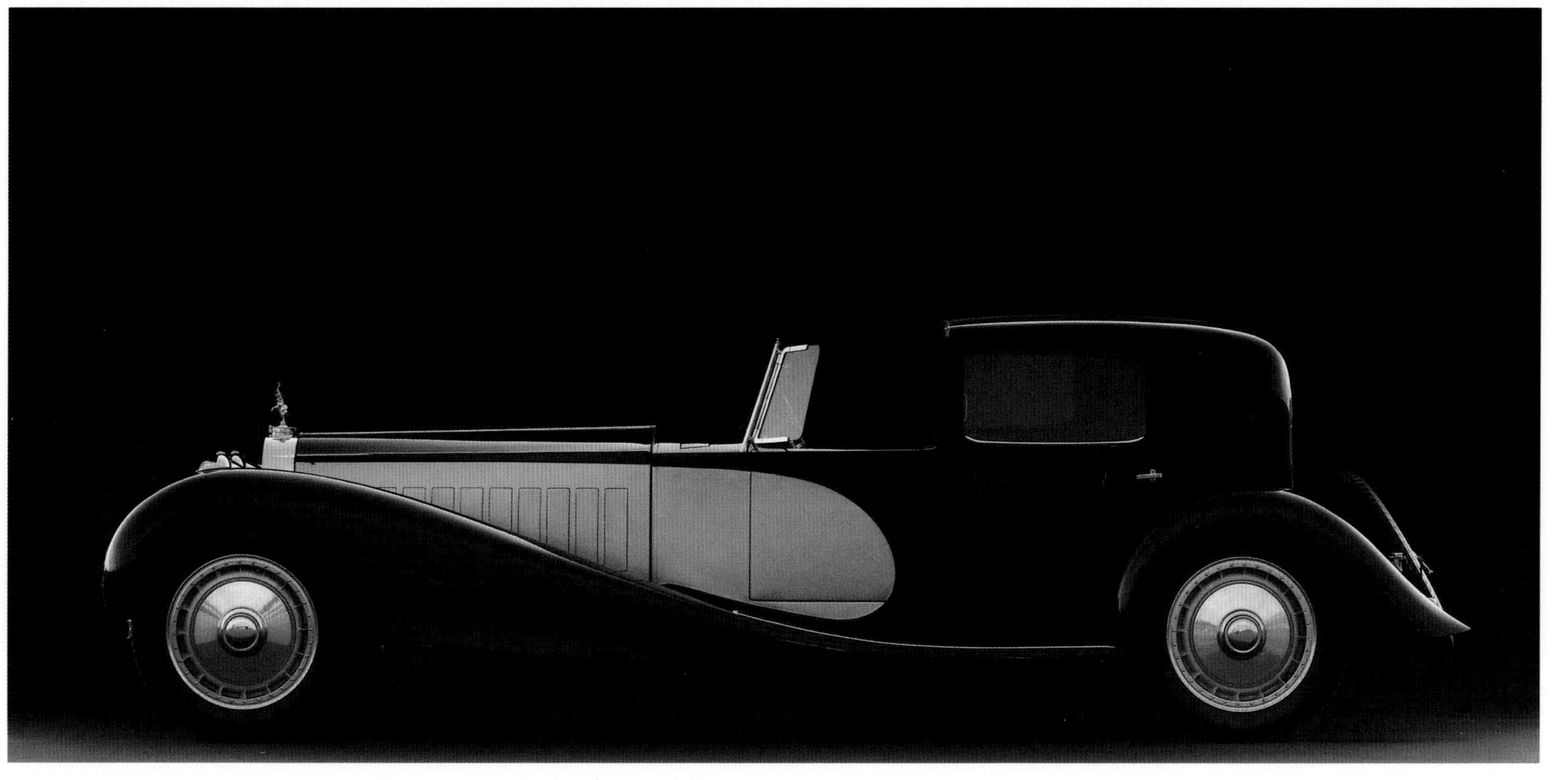

Type 37A derived from a racing heritage. Swift, and blessed with precise handling and surprisingly good braking for a Bugatti, it carried one or two occupants in glorious discomfort and the sustaining illusion of being always in the lead at Le Mans. The excellent engineering was exactly matched by the car's economy of line—and by its annoyances. In some magical way, however, owners felt it was better to have a car that was uncomfortable, difficult to maintain, and expensive to repair. In fact, some of this auto-masochism lingers today in our Ferraris, Lamborghinis, and Lotuses, does it not?

Bugatti's technical excellence was sometimes achieved with utter luxury, as in the case of the fabled Royale, a 300-horsepower giant that recently fetched £5.2/$9.8 million at an auction. Only six examples were ever built of this huge car that was intended to put Rolls-Royce in its place once and for all.

France was also beautifully served by other sports cars than the Bugattis and by production vehicles as well. French *carrossiers* flaunted the French curve with abandon, and the Delahayes, Lago-Talbots, and Bugatti 57Cs were given opulent, singing lines that must have maddened the metal-benders as much as it made the wealthy owners delirious.

At an equal but more subdued level of affluence, the lines tended to follow formal English and American practice, softened by the characteristic French elegance. The Hispano-Suiza reigned supreme at the top of the pecking order, but lovely limousines and towncars were built by Panhard and Renault; and there was also the BUC, a front-wheel-drive L-29ish creation of the Bucciali freres.

Once the veil of aristocracy was lifted, the French were treated again to flowing curves in such cars as the Peugeot 401, 601, and 402. The 601 featured an all-steel retractable convertible top in 1936, while the 402 managed to look like a vastly refined Airflow.

The Citroën Traction Avant was more deceptive, dressed in an exterior shell not unlike that of a 1934 Ford. The car's tasteful but uninspired bodywork concealed its advanced unitary construction, front-wheel drive, rack-and-pinion steering, torsion bars, and other modern elements, which kept it in production for twenty-three years (until 1957). A favorite car of French gangsters for its speed and road-holding qualities, the Traction Avant of the 1930s has been called by some aficionados the car of the decade, and it is gaining in value as a collectible.

Other French midrange products of the time, however, tended to cloak conventional mechan-

PREVIOUS PAGE:
Oddly enough, a long straight eight engine provided coachbuilders with the opportunity for lots of curves, as this 1938 Delage D8/120 Aero Coupe shows *(top)*. This car received a special body from Letourneur et Marchand, and was exhibited at the 1939 New York World's Fair.

Ettore Bugatti had the inspiration to build a car for royalty *(bottom)*. If things had gone his way, the Windsors, Habsburgs, Hohenzollerns, and Romanovs, among others, would have queued up to purchase the Royale. The car was rebodied to this configuration for King Carol of Rumania and reportedly survived the war crated in a sewer in Paris. With a wheelbase of 170 inches, this 1931 Type 41 Coupe de Ville was powered by an 8-cylinder, 300-horsepower engine, and came with an unbeatable warranty: Bugatti himself would maintain the car, free, forever.

OPPOSITE:
A toy Bugatti Type 35 lies abandoned as the real thing races by in the French Grand Prix, August 1929. *(Photograph by J. H. Lartigue)*

ical arrangements beneath conventional body work. These included the standard products from Renault, Hotchkiss, and Berliet.

At the lower end of the scale, the masses (a misnomer, of course, for even small cars were only within the means of the middle class) had available the tiny Citroën precursors of the 2CV, and the Simca version of the Fiat Balilla. These cars succeeded better than did their English and American counterparts in preserving attractive proportions within smaller dimensions. American small cars—in particular, the early Bantams—tended to look like comic strip cars, the owners bulging disproportionately out of them; in contrast, perhaps because the owners themselves were typically of smaller stature, the French cars looked correct in scale.

Just as the Bugatti was quintessentially Gallic, so were the great Mercedes cars of the late 1920s and 1930s uncompromisingly Teutonic. The Mercedes SSKs, like the Bugattis, had an aviation ancestry, descending from the engines that pow-

ered the Albatros, Fokker, and Pfalz fighters of World War I. Ferdinand Porsche left the Austro-Daimler works for Mercedes-Benz, to succeed Paul Daimler, who was going to Horch, whose founder (and the founder of Audi), August Horch, had just departed.

Porsche stretched out (but did not sufficiently strengthen) the Paul Daimler–designed heavy frame so that it flexed its way over the highway. The long hood accommodated the huge six-cylinder in-line engine that he fitted with a twin-blade supercharger. Unlike the Duesenberg supercharger, which operated constantly, the Mercedes supercharger was cut in as required. The SSK's primitive suspension system featured a live rear end, stiff springs, and a solid front axle. It suffered from an incongruously poor braking system.

The loss of World War I, the subsequent rise of inflation, and the worldwide depression that followed constrained the production of cars in Germany. An adequate supply of Mercedes-Benz sedans remained available for the wealthy. And as in England, where the Bentley served as an alternative to the Rolls, in Germany the Horch sufficed as an alternative to the Mercedes. A more rarefied limousine could be obtained in the Maybach Zeppelin, while the BMW 328 catered to sporting instincts.

Italy is justly praised for its art, music, food, and wine, for the Italians are a sensitive people, capable of appreciating the best things in life. The appreciation of a classic car, like the appreciation of a fine wine, is a task for all the senses; and Nicola Romeo excelled with his Alfa Romeo series. The Alfa's lines provided beauty for the eye, while the sense of touch was gratified by the car's precise handling. The Alfa's engine noises (especially those of Vittorio Jano's 6C 1750 Gran Sport) were intoxicating, from a deep-throated idling rumble to a supercharged scream at high speed when driver Enzo Ferrari hurled it down the highway. And the smell, of fine leather and partially burnt castor oil fumes (used to lubricate the supercharger, just as you might mix oil and gas in a lawnmower), was a combination never to be forgotten.

Joel Wolf "Babe" Barnato —the man whom W. O. Bentley said was "the best driver we ever had . . . the only driver who never made a mistake"— sketched the original design of this 1929 Bentley Speed Six Coupe on an envelope. It became famous for doing what most other cars of the time claimed: beat the Blue Train from Paris to Calais.

GJ 3811

Just as the nickname ''Bug'' was perfect for the Volkswagen, ''Topolino'' was a match for the 1936 Fiat. With its thirteen-horsepower engine, it could run on Mussolini's *autostrada* with the best of them.

Lovers of Italian sports cars are a breed different from any other type, alternately belligerent and defensive, and this may be either a cause or an effect of the cars themselves. Wonderful to drive, capricious to live with, they demand affection as well as maintenance. Mussolini, besides pursuing his well-known interest in the railways, also forced the creation of the *autostrade,* linking major cities with roads that Caesar would have been proud of and that Hitler emulated. These were perfect for hot little two-seaters to compete upon. The Alfa Romeo 6C 1750 Gran Sport was capable of 95 mph on these highways, slaloming past the tiny 40-mph Fiats.

Gran Sports were racing cars that had been detuned to approximate touring standards and cloaked in spartan, lightweight bodies of great elegance. At a time when the country was seeking heroes everywhere—Marshal Italo Balboa in the air, Umberto Nobile over the Pole, Pietro Bagdolio in Africa—Italy found many in its own back streets. Tazio Nuvolari, Giuseppe Campari, Achille Varsi, and others became legends driving the most refined of a very refined breed of Alfas in the Mille Miglia.

The 1750 Gran Sport was developed and honed into the lovely 8C 2300 of 1933. In its most potent form, power was supplied by a supercharged double-overhead-camshaft straight engine delivering 165 horsepower to a lightweight chassis that weighed just over 2,000 pounds. The bodies were streamlined, and the car turned heads during the five years of its production. Expensive, at the equivalent of about $10,000 in United States currency at the time, it laid the groundwork for the Alfa 6C 2500, which was built from 1939 through 1943. It represented the pinnacle of the less than 13,000 cars the firm produced during its first thirty years of existence, and it pointed the way to Alfa's first postwar designs.

Low production was the rule rather than the exception in Italy, even for such illustrious makes as Isotta-Fraschini and Lancia. The dominant manufacturer had always been Fiat. A huge corporation, with interests in virtually every industry, Fiat sought to make cars that were easy to produce and easy to sell. In the prewar years, Fiat styling was essentially American, with the happy exception of the Topolino (little mouse).

Franco Fessia, the Topolino's designer, was able to achieve perfect proportions in this runabout, which could achieve a 51-mpg fuel economy that almost matched its 52-mph top speed. The car, while conventional mechanically, was exceedingly well-conceived. Windows slid to open, rather than rolling up and down, to narrow the doors and maximize the interior space. A folding top rolled down on rails to let in the Italian sun. The entire front portion of the hood moved forward to reveal the tiny 13-horsepower four-cylinder engine, behind which was mounted the radiator (as in the L-29 Cord). This efficient Fiat 500 appeared in 1936 and was built until 1948. Intended at first to be a two-seater, it proved to be so popular that space for shoe-horning two additional (very small) people in had to be provided. The Topolino handled beautifully, and its wonderful fuel economy was even more important in Italy in 1936 than it would be today. Sadly, when Fiat later cloaked essentially the same mechanicals in a more modern sheet-metal body, the car lost much of its appeal.

Great Britain had many great cars in many categories, all zealously championed by vocal adherents. The one car that looms in British automotive history as the Duesenberg does in American

Twenty-six-year-old MGM leading man Robert Montgomery perched on an Austin, just after finishing the unforgettable film *War Nurse*.

automotive history, however, is the 1930 8-litre supercharged Bentley.

W. O. Bentley began his career as a locomotive engineer, but soon began designing engines. He is responsible for creating perhaps the best of the aircraft rotary engines, the B.R. 1 and B.R. 2, which powered Sopwith Camel and Snipe fighters in World War I. After the war, he built a series of 3.0-litre cars that established an enviable reputation for speed and reliability. He liked to build his engines overstrong and understressed, depending on their cubic inches to generate the power he required. Bentleys ran successfully in the Le Mans race, taking fourth in 1923 and first in 1924; and Ettore Bugatti is supposed to have murmured that Bentley built "the fastest lorries in the world."

Although beset with financial problems stemming from a development race with Rolls-Royce,

CALIFORNIA
933 J12

Bentley produced cars that won the 1927 and 1928 races at Le Mans. But other cars were proving lighter and faster, so W. O. Bentley was prevailed upon (against his instinct) to supercharge a 4½-litre engine for the 1930 race. The result was a gorgeous two-seater with an open body and a four-speed gearbox. Its maximum speed was 125 mph, fast enough for it to place second in the race.

Unfortunately, producing a sufficient number of "blower Bentleys" to qualify for the race was the financial strain that forced the firm into receivership. W. O.'s instinct had been right: it wasn't a good idea to overstress his engines. Rolls-Royce extended a gracious compliment by acquiring the firm and established Bentley as its understated second line.

Rolls was of course the premier luxury car of the period, although the rich could also choose from among handsome Daimlers, Armstrong-Siddleys, Lagondas, Jaguars, and even jumped-up Vauxhalls and Humbers. The beauty of these cars lay in their opulence rather than in their lines. Dignified and comfortable, such cars were intended by the designers to pamper the owner looking out rather than impress the passerby looking on.

The caste system hung heavily in England, and what you drove made a precise statement about who you were. Thus, a tradesman in a Rolls was droll, while a lord in a Vauxhall was impoverished. The standard among Britishers was to select the proper car, just understated, for one's rank. The car of the royal family was a Daimler, but a duke could drive a Rolls. A person who ran an aircraft company could drive a Rolls, but a Bentley was in better taste.

Just below this rarefied (and artificial) level, there were no holds barred. Reid Railton had worked as an assistant engineer at Leyland Motors before beginning a career of fine-tuning the performance of a wide variety of English and American components. Railton, the English equivalent of Frank Curtis, completely upset the association of caste and price with wonderful, swift-paced hybrids. One of his most successful ideas was to combine an utterly distinguished English exterior—Bentleyesque, in fact—around a Hudson chassis and engine! It was faster than most of the competition and more reliable than all of it, and it cost only a quarter of what a Bentley cost. Why Hudson didn't make more use of it in its United States advertising is a mystery.

The 1930 Hispano-Suiza J12 Faux Convertible, considered by many to be the finest car of the classic era. Hispano-Suiza typically combined advanced engineering with exquisitely simple design. The engine was a huge 9¼-litre V-12; its four-wheel braking system was adopted by Rolls-Royce and used until 1967. Only 120 of the J12 chassis were made.

The English Baroque was graced by many other cars of note. William Riley founded a firm near Coventry in 1898 and began producing a series of highly desirable cars (the Rileys) bred on the race course. Humber followed a different route, working itself up from the Chummy of 1923 to lovely Pullman limousines. Alvis, Sunbeam, Rover, and Singer all built fine middle-class cars. Styling faults of proportion, line, and harmony, however, left these British cars—while different from the type produced by American practice—no better.

Nor was the common man neglected. Sir Herbert Austin worked his way up from a Yorkshire farm to become an absolute autocrat of motordom. In 1922 he introduced the Austin 7, designed jointly by Austin and by 18-year-old Stanley Edge. It had a four-cylinder engine, hydraulic brakes, and the key to the working Briton's heart, despite its almost cartoonlike proportions. In the next seventeen years, 290,000 7s were sold to people who might otherwise never have owned a car. It sold in many countries, including the United States, and laid the foundation for an entire series of larger cars.

Ultimately, however, every Englishman saw himself as being cast in one of two molds. Either he was Bertie Wooster, tootling off to Blandings Castle in the old two-seater, or he was a proper English Nuvolari, driving as relentlessly as that famed Italian racer, but drinking sensible ale rather than Chianti in raffia bottles. The psychological demand for M.G., Morgan, Frazer-Nash, Alvis, and Riley sportscars was incessant, even though the economic demand was relatively slight. These were cars that could be coddled in a garage through the week, and then brought out to tear around the countryside on the weekend.

The minimal bodies of these sports cars purged the standard English styling of its defects, leaving behind the slashing good looks that set American

Dutch Darrin did as much as any man to keep Kaiser-Frazer from going under. His Kaiser Darrin fiberglass-bodied vehicle was handsome—and radical (the doors slid forward into the front fenders). Less than seventy cars were built in the original run. ***(Photograph by Curtice Taylor)***

hearts aflame and continue in production to this day, either in the Morgan or in the replicas created to honor the M.G. and the Frazer-Nash. Drivers devoted to these lovely cars could have borne the danger and discomfort of World War II with equanimity if they had only known that their reward was coming. It would be the Jaguar XK-120.

American styling persisted in other countries where variations could have developed economically, simply because cars were produced in limited quantities. Volvo maintained a steady line in Sweden, building cars of wandering American influence; sometimes they appeared to be Chryslers, and other times Fords. Both Toyota and Nissan were making cars of a sort in Japan. Their logic seemed to be that, if all those American cars looked a certain way, theirs should, too.

THE ESCAPIST PERIOD (1946–1976)

New 1946 cars were available on the American market before Japan officially surrendered. The automobile industry geared up for civilian production with even greater enthusiasm than it had shown in its remarkable dedication to war work. Some materials were in short supply, and the distribution network had to be reestablished, but the demand for new cars was unlimited. Styling was not to be a major consideration except in the case of new entries into the arena.

Established manufacturers reinstated 1942 models into production with minor facelifts. Both Kaiser-Frazer and Tucker (totally new names) wanted to get off to a head start and become established by the time the sellers' market ended. Similarly, older firms for whom the bell had begun tolling—Hudson, Nash, Packard, and Studebaker—saw the Big Three's stylistic stagnation as an opportunity to improve their market share.

Kaiser-Frazer seemed to present the greatest prospect of success, introducing brand new envelope bodies and pioneering the concept of sumptuous, sometimes exotic interiors. The chosen engine was the reliable, efficient Continental of 100 horsepower. Although they were comparatively expensive, some 325,000 cars were produced in the firm's first two years. When the sellers' market began to dry up in 1948, sales plummeted, with the senior line (Frazer) being most severely hurt.

Strapped for funds, the firm managed to prolong its existence with a brilliant styling tour de force, Howard "Dutch" Darrin's 1951 Kaiser. Had the company been able to acquire a V-8 engine to place in this handsome package, it might be in business today. Without it, the car simply didn't have the snap of the Oldsmobiles and Buicks in its price range, and neither the continuation of excellent styling nor the addition of a Graham-type supercharger could prevent the firm's demise in 1955. Argentina continued production of the same basic car as the Carabella for another seven years.

By 1949 there were new rules to the new car game. Buyers were demanding both the styling and the horsepower that GM had taught them to want. In an earlier decade, the Kaiser's incomparably beautiful lines would have sufficed, but it could not withstand the new marketing methods. The customers wanted tire-peeling acceleration, and this meant bigger, more expensive engines (which, of course, were easier for the larger companies to deliver first). The ante had been raised. Now, not only did the independents have to deal with annual styling changes and complete redesigns every two or three years, they had to match the Big Three's ever bigger and more powerful sets of engines.

Studebaker was felled by this new complexity, despite having been "The First by Far With a Postwar Car," according to its advertisements. Regardless of their disadvantages, Raymond Loewy and the Studebaker team twice created landmark designs. The first were the "coming or going" cars that attained immediate celebrity (verified by appearing in a Bob Hope monologue) upon their introduction in 1947. In this instance, Virgil M. Exner was very influential in the styling. The fore and aft sections of the car were about equivalent in size and shape, and the extravagant Starlight coupe featured a futuristic four-section wraparound rear window. In contrast to Nash, Packard, and Hudson, all of which introduced new bodies the following year, the Studebaker wore its styling well until a replacement model arrived in 1953.

This time Loewy teamed with Bob Bourke to create what many people believe is the most beautiful American production car of all time. Its sculptured, understated refinement ran in direct opposition to the styling of its competition. The basic body would linger on until the 1960 Hawks, its face lifted and its fenders tweaked, but its lineage still identi-

Detroit, 1953. ***(Photograph by Harry Callahan)***

fiable from the benchmark of seven years before. Studebaker also built excellent V-8 engines, but its designs and the cycle of its engine were usually out of step, and the predictable fall in production began that eventually led it out of the market, notwithstanding its morganatic marriage to Packard.

Hudson and Nash, at first individually, and then as a merged unit, failed on both counts—styling and engines. George Romney's Rambler breathed life into the new American Motors, and the Jeep lines of the later merger with Willys kept things going for a bit longer. Eventually, however, years of losses drove it first into a joint relationship with Renault and then into the arms of Chrysler, whose major interest in the acquisition stemmed not from the American Motors line itself, but from the Jeep four-wheel-drive family of vehicles.

The first postwar cars were handsome, well-balanced vehicles in the modern idiom, but with typical American features of flow-through fenders, wide, well-chromed grilles, and embellished taillights.

Then a kind of runaway pituitary growth madness set in, and stylists with long, flowing arms and

BELOW:
The 1956 Lincoln Continental Mark II was a far greater styling success than its makers appreciated.
OPPOSITE:
The French carried Art Deco into the middle of the twentieth century with the fabulous Henri Chapron coachwork of this 1948 Delahaye Type 135MS.

DELAHAYE

THE EVENING SUN
U.S. Legation Staff Hurriedly Leaves Finland
20 BIG NAZI PLANES DOWNED
New Tax Bill Provides Abatements On Sliding Scale
Flashes
Majority Of U.S. Legation Quits Helsinki
Entire Convoy Destroyed Over Gulf Of Tunis
British Concentrate Fire On Hill Fortress
Featherbedding By U.S. On Railway Charged
No Justice In Japan, Says Scribe Tried There
Prison Break 'Hilarious' Georgia Probe Shows
3-31 MARYLAND 1944
187-187

PAGE 172:
Baltimore, 1943. A hard-hatted Baltimorean standing protectively in front of his 1937 Plymouth. The gas shortage of the mid-1970s was child's play compared to the shortages during World War II. *(Photograph by Arthur Siegel)*
PAGE 173:
The unmistakable looks of Dior and Buick—Paris, 1953. *(Photograph by Willy Maywald)*
PAGES 174–175:
Close—but not quite: a 1959 Cadillac Coupe de Ville *(left)* meets a 1959 Cadillac Fleetwood *(right)* with its additional soupçon of chromium trim.
LEFT:
Shades of Jack Kerouac; US 90 en route to Del Rio, Texas, 1956–57. *(Photograph by Robert Frank)*

flexible elbows seemed to step in and dictate what American cars should be. Engineering changes persisted, but they were subordinated to the whims of the stylist to a degree never before tolerated. The styling changes were incredibly stupid. Cars got bigger and heavier, while their interiors either stayed the same or grew smaller and less comfortable. Enormous overhangs developed at the front and rear of cars. A foot of empty space often separated the chrome-toothed grille and the actual radiator, while in the rear a cavernous trunk yawned behind a 1-foot-high lip.

It was extravagance without conscience, vulgarity without vision, statement without content. No commensurate effort was made to keep suspension capabilities comparable to horsepower increases, beyond what was necessary to soak up and deaden bumps from seams of tar in a concrete roadway. Detroit meant for cars to go fast or to go straight; a luckless owner who wanted a car to do both was in for trouble.

The nation was in an economic upswing of unprecedented scope and duration; and bad though the cars were, we bought more and more of them. As volume went up, quality control went down at every level. It was not uncommon to see a Cadillac or a Lincoln with mismatched trim lines, doors clearly not fitted correctly, gaps in assembly points, and wretched painted surfaces. Repairs were constantly required and constantly more expensive. Tires wore out (often within 10,000 miles) from soft and mushy front suspensions that could not be kept in line. Even rear suspensions got out of line.

There were almost as many claimants for the title of ''Most Hideous'' during the Escapist period as there were claimants for the title of ''Most Beautiful'' in the 1930s. The Edsel is always picked on as the 1950s' most egregious example of corporate error, but the contemporary GM products were no better. The 1959 Lincolns were stunningly awful—as if the French airplane makers of the 1930s had designed them—although many critics say that the 1958 Olds was fully as atrocious.

Chrysler Corporation provided a variation on the theme, creating uglier cars within the universe of ugly cars. It had an endearing (if costly) habit of first creating its own proprietary styling disaster, and then dropping the design in embarrassment on a tide of public disapproval. Finally, it would bring out (two years late) a GM-like car that was even worse.

It is difficult today to understand how anyone in management could have permitted the endless, mindless excesses to continue. But ironically, management thought it was doing wonderfully well, as it focused unblinkingly on the bottom line. Profit could be manipulated by training all engineering attention on the objective of shaving a few cents per part. When you produce a million cars, a few cents savings on each results in hundreds of thousands of dollars of pure profit. Why fix it if it ain't broke? The truth was, however, that it was terribly broken—perhaps beyond repair—but that the break would not be apparent to the Big Three until Japan pointed it out with endless exports. The top salaries in the country were earned by a small group of men who collectively could not create a single car that was not a wasteful affront to styling.

Such generalizations are never entirely true, of course. A few good-looking cars were made during the period, usually smaller ones whose physical dimensions and fiscal considerations stayed the designers' extravagance. The classic is the 1955 Ford Thunderbird, a two-seater easily ten times as beautiful as the four-seater that succeeded it in 1958. The two-seat Thunderbird, with its low lines, racy hood, and minimum of decoration, is still a head-turning car that, in first-class condition today, commands a $20,000 price. The 1958 Ford was only 18 inches longer and 1,000 pounds heavier than its predecessor, but it looked twice as big. Chrome was lavished front, side, and rear, and no harmony at all existed between the styling of the body and the styling of its top. A first-class example will bring only $7,000 today. And yet the ugly 1958 model sold 50 percent better than had the 1955 original—a statistic that strongly suggests why the pattern of design excess was followed with such ardor by American manufacturers.

Ford repeated the triumph in 1964 with the Mustang, produced when Lee Iacocca ordered masterful plastic surgery to be done on the prosaic Falcon. Similar in size and lines to the original Thunderbird, with a long hood and a short, square rear, the Mustang combined genuinely sporty looks with a selection of luxury and performance options that would satisfy anyone. A plain-jane Mustang with a small six-cylinder engine had a base price of under $2,400. Or you could load it up with a high-performance V-8 and every option in the book, from air conditioning to remote control day/nite mirror. It was a tremendous hit, and inspired some of the other small, attractive cars of the period, including the Chevrolet Camaro and the Plymouth Barracuda.

Other stylish cars of the period include the notorious Corvair (its styling soon to be aped by BMW), the 1961 Lincoln (which, like an overweight movie star back from the fat farm, somehow kept its large proportions contained within comely lines), and the 1962 Buick Riviera. The Riviera showed what GM stylists could do in a moment of thoughtful restraint: it was so understated that its bulk was minimized, and its lines were taut and cohesive.

The King of Rock and the King of Cars: Elvis Presley with one of the hundreds of cars he owned—a 1956 Lincoln Continental.

Every car has its adherents, but the truth is that most other American cars of the period were stylistic schlock, incapable either of pleasing the eye or soothing the nerves.

The tendency toward garish styling was undoubtedly worldwide, but Europe and Japan managed to avoid the worst excesses because of their automotive industries' different economic situation. In practical terms, America's foreign com-

Cited by David Halberstam in *The Reckoning* as the ideal Japanese car company with which to compare the follies of American manufacture, Nissan has since had its full measure of corporate problems. The 1988 Maxima *(left)* offers surprising performance and value, however, while the Nissan Pulsar NX SE *(right)* is a product of the marketing technique that both Nissan and Toyota have used with great success. When originally introduced in 1982, the Pulsar was an entry-level econobox. Each year, it has been improved and upgraded, ultimately blossoming forth in its sporty, NX SE mode.

OPPOSITE: A small car with a low sticker price—the Festiva *(top)*—has done very well as it has ridden the rising tide of Ford's expectations. The flip side of this situation is illustrated by General Motors' response to Mercedes Benz. Despite its sporty lines and the cachet of the name, the 1987 Cadillac Allante *(bottom)* has received only a lukewarm reception.

petitors did not have the luxury of doing the foolish things done here. Tradition reinforced constraints unknown in the United States. Roads were narrow, parking was impossible, automobile taxes were based on horsepower, and fuel taxes were high—all forces tending to keep cars smaller and more functional. Perhaps the most decisive factor was the canny European demand for quality in a purchase of the magnitude of an automobile. Names that had sounded strange to American ears—Saab, Volvo, Volkswagen, Porsche—now became not only familiar, but fashionable. The master of sales, Chevrolet, even deigned to design a rear-engine competitor (which grew into the Corvair) to counter the appeal of the VW.

But changes were coming. An insidious trend was clearly discernible in the premier lines of many European makes toward American-sized, American-styled cars. The Austin Westminster, Vauxhall Cresta, Rover 2000, Borgward Isabella, Ford Zodiac, Simca 1500, Volvo Amazon, and even Toyota Tiara were larger than their earlier brethren, and traded heavily on the size "advantage" in their advertising as well.

The European oil crisis following the Suez diplomatic follies of 1956 nipped this trend in the bud. Perhaps the most unexpected result was that the path to the automotive future was pointed out by a car that had been regarded by many to be just a hyperthyroid bubble car, a freak, a passing fancy: Alec Issigonis's masterful miniature, the Morris Mini-Minor.

The Relearning Period (1977 to the Present)

In an age of skyrocketing sticker prices, the idea that car manufacturers and consumers are relearning the art of economy may at first sound absurd. But there are many economies other than purchase price. The driving force behind the modern movement toward economy has been the desire to economize on fuel use. But the present age seeks economies in many other areas as well. The current demand for improved quality—so great that even Detroit is responding well to it—is clearly a demand for long-term economy. At issue here are savings in the use of materials and in the use of space. The most important economy, however, is the saving of life and limb achieved by making better brakes, stronger crash-resistant structures, and protective safety belts, air bags, and the like.

Alec Issigonis may have had some of these larger economies in mind when he created the Morris Mini-Minor, but it is more probable that he was simply providing his usual innovative engineering response to a tremendous and more immediate challenge. The slender, aristocratic Issigonis dreamed, conceived, and created as an artist. Each of his automobiles was a balanced, harmonious entity derived from his engineering training and his hands-on experience.

His first great postwar success, the Morris Minor, illustrates his modus operandi. Working

with a small team, Issigonis created an attractive small car, with torsion-bar independent front suspension and rack-and-pinion steering. Very late in the design process, he decided that the car looked too narrow for its height, so he had the prototype sawed in half longitudinally and moved the two halves apart until they reached a width that he thought looked right. The resulting 4-inch gap was filled in with sheet metal and structure, and it conferred upon the car not only a handsome appearance, but better handling characteristics!

Extraordinarily advanced at a time when cars didn't need to be to sell, the Minor debuted in 1948 and was an immediate and sustained success. In production for more than twenty-five years, the Minor was built in numbers exceeding 1,500,000; many are still cherished (and driven) in England and the United States.

In 1952, Morris and Austin merged to form the British Motor Corporation, the same year that Issigonis left to join Alvis. He returned in 1956 and was given a clean sheet of paper on which to create a small car that would respond to concerns raised during the fuel crisis occasioned by the abortive 1956 Suez operation. (Issigonis was a "clean sheet of paper" expert, accustomed to sketching out ideas on a scrap of cardboard and handing them to a staff member to execute for manufacturing purposes.) Using the courage born of long experience, he created a front-wheel-drive car with the engine mounted transversely. The resulting great space savings were translated into passenger accommodations, yielding a car only 3 m/10 feet long but still capable of seating four persons.

The Mini-Minor was a good-looking, good-handling car. It set the engineering style for the future, for today almost every front-wheel-drive car in production essentially follows Issigonis's formula. But it is doubtful whether even Issigonis was aware of the revolution he was creating, since it took the introduction of a challenging style of car from another quarter and another culture to make clear the full importance of the Mini.

QUALITY EMERGES AS A STYLE DETERMINANT

The Volkswagen—the first messenger of German quality to make a genuine impression on the buying public—had prepared the way. New Volkswagen owners used to get an intense personal pleasure out of the fact that their Bug was so tightly built that it was difficult to close the front door without cracking a window. And of course the Volkswagen advertisements were masterful, capitalizing on the wholesome plainness of the Bug's appearance, as well as on its quality. Film clips of VWs floating merrily away on a raging river, briskly starting under mounds of snow, or roaring past overheated liquid-cooled cars all contributed to an indelible impression of quality.

This impression was reinforced by the offerings of Porsche, Audi, BMW, and Mercedes-Benz. There were far fewer of these cars imported, but the prior inoculation with Bug fever made America and the world far more sensitive to what they represented. The marketing position of these makes was also helped by the general state of prosperity in the industry. In boom times, an expensive price tag can almost be an advantage. Long before there were yuppies, there were car buyers who saw that they could indulge themselves with a car they wanted and have the social status conferred compensate for the price differential.

Mercedes-Benz took advantage of this phenomenon to dismantle General Motors' monopoly on styling. The first signs that this was occurring could be seen in the wistful, buck-toothed Mercedes-like grilles that appeared on the Studebakers and pseudo-Packards of the time. There was a marketing connection, of course, but the greater significance of the styling trend was that Studebaker management and the buying public considered a Mercedes-like grille on a Lark to be an advantage.

It was not long before Mercedes-Benz-inspired lines began to appear in almost every make. The Teutonic square body with defined hood, prominent chromium grille, and cubical trunk was seen almost everywhere. By the 1980s, most of the larger Big Three cars seemed to be engaged in a Mercedes look-alike competition. Fords that threatened to come apart when the door was slammed nonetheless had a square, Mercedes look, as did Chrysler products. General Motors was caught in the same trap, although Pontiac—which for many years had a split-V grille as its signature—somehow found itself looking like a BMW imitating a Mercedes.

The tremendously important point that the Mercedes style was fashionable because the Mercedes was a *high-quality product* was missed. Yet a funny thing happened to the Mercedes-look phenomenon on its way to the Orient: Honda adapted the style to its early Accord four-door sedans, and within a few years many makes chose to look more like a Honda than like its German inspiration. Honda had faithfully followed the Mercedes example by establishing a reputation for extraordinary quality that in turn made its styling coveted.

This change—quality imparting desirability to styling—was Japan's secret for its incredible incursion into the world's automotive markets. The country whose products in prewar years had been synonymous with junk was suddenly bringing excellent small cars into the marketplace. The fact was ignored by Detroit, even as it was applauded and accelerated by the American buyer. Detroit's attempt to meet the challenge with a "small big car" failed miserably. Pintos and Vegas sold in large numbers, but they were poorly made cars that reinforced, rather than countered, the Japanese idea. For the combination of German and Japanese quality had established in the public's mind (perhaps for the first time in fifty years) a prevailing attitude that quality was more important than how the sheet metal was bent. (The struggle did provide a niche for the Gremlin and other small cars from American Motors, which, if they had attained the necessary quality, might have improved the company's market share significantly.)

Curiously, the single most important car in the industrial firestorm that broke upon American shores had more of the Jaguar XK-120 than of the Mercedes about it. The Datsun 240Z (was it any accident that the number chosen was twice 120?) swept aside all precedents and brought to the Japanese automobile industry a previously unattainable prestige and status.

The 240Z overwhelmed the car-buying public as the Jaguar XK-120, the original Thunderbird, and the Mustang had, for it offered advanced styling and genuinely high performance at a very reasonable price. Buyers queued up at the dealers, and a black market quickly arose to serve buyers too impatient to wait for delivery.

The advent of small, stylish, fuel-efficient, high-performance Japanese cars shook Ford from its internecine bickering, General Motors from its incredible profits and fatuous complacency, and Chrysler from its state of near-bankruptcy to prosperity. The shock waves jolted even British labor unions, and the "I'm Alright, Jack" syndrome that had utterly demolished Jaguar's reputation was turned around—the first step toward rehabilitating British car-making. The introduction of the Anglo-Japanese Sterling is the most recent expression of this beneficial trend.

Never properly appreciated in the United States, the Morris Minor is one of Britain's great cars, where survivors of the 176,000 built over its five-year production run are lovingly cared for by loyal owners.

LEFT:
The Datsun 240Z was one of the greatest automotive coups of all time, taking the American public by storm with its sleek good looks, sporty performance, and relatively low price. Over the years the Z series has lost its original lean elegance, but it is still among the most popular sporty cars around.
RIGHT:
When the first Hondas appeared, they were regarded as also-rans destined to be trounced by Detroit's corresponding products. But each succeeding wave brought new and better models, and Honda's image soon changed from that of toy to being a synonym for quality. The 1988 Prelude Si 4WS is the world's first production vehicle to use four-wheel steering.

The net result, in terms of styling, has been for the most part one of healthful restraint. The originators continue to be most successful with the trend. Mercedes, Honda, and Toyota maintain the style and quality that most manufacturers seek. Efforts to improve drag reduction happily fostered the softening of the original square Mercedes look to the present attractive, smoothly rounded appearance of Accords, Camrys, and 300Es, as well as of 626s, 6000Es, Galants, and many others. Volvos have remained square but manage a Scandinavian coolness that sells well.

Firms that depart from the formula meet with different degrees of success. Audi did very well with its low-drag aerodynamic 5000 series. Saab changed its DKW two-stroke engine image to one of a high-performance tank, making basically unattractive lines somehow comely and convincing. Jack Telnack encouraged Ford to embark on its greatest gamble since the Edsel by effectively adapting the aerodynamic lines of the Audi to its own distinctively different formula. Its Sable and Taurus lines met with instantaneous approval, catapulting it into the unquestioned styling lead among the American Big Three. Telnack's brave direction moved in a blazingly swift period from unprecedented departure to standard setting. But it is essential to recognize that the Ford styling achievement would have gone for nought, had there not been a comparable improvement in quality. Ford has recognized this itself in the redesign of its flagship Lincoln Continental with the Telnack style. The two elements—quality and style—have resulted in Ford's becoming the most profitable American manufacturer, and have given it the courage and the wherewithal to acquire Aston Martin and Lotus.

Iacocca has worked miracles with the basic K car design and engineering, producing a progressively differentiated line of successful cars that are a vast improvement over the previous Chrysler cars. The number of variations on a basic theme has been remarkable, and the far-seeing association with Maserati and the acquisition of Lamborghini speak well for the future. Chrysler has shown equal perceptiveness in its new advertising campaigns.

On the wrong side of the tracks stands poor old General Motors. Despite its concessions to the Mercedes look, the firm's styling efforts of the last decade have been abominable, apparently reflecting a corporate desire to retain the familiarity of the line as the cars were down-sized to meet the realities of the modern fuel situation. The result has been a series of hybrids—cars that looked strangely out of proportion, with trunks somehow telescoped into bodies too large or too small for them, as if there had been a mix-up at the Fleetwood and Chevette production lines. Coupled with the styling anomalies were continuing quality-control difficulties that often led to model recalls of massive proportions.

Old ideas, old plants, excessive layers of acquiescent management, and an in-bred contempt for public taste prolonged GM's resistance to making desperately needed changes for an unconscionably long time. The one signal that had in the past

always ruled at General Motors headquarters, decline in percentage of market share, brought forth only the palliative responses of rebates and improved financing. The need to reach down deep within its enormous resources and pull out an entirely new approach to car-making seemed for years to be beyond GM's capabilities.

After an agonizing decade, the worst is obviously over for GM, in terms of both styling and quality. The overall proportions of General Motors cars (particularly the larger ones) are beginning to take on a reasonably attractive form, and slowly but perceptibly the quality is improving. Cadillac, its image badly tarnished by quality problems and by the fact that in the past its body types were shared by lesser GM marques, is once again going to be built exclusively in its own factory. Chevrolet has allied itself with Toyota and is producing the Corolla under the Nova name plate. Pontiac is stressing its BMWness with the 6000, and both Buick and Olds are making similar efforts.

During the dark period of Detroit's travail, things have not gone all that well in Kenosha, Wisconsin, either. American Motors has never had the finances nor the design and engineering ability to produce a car of sufficient quality and styling to stem its decline toward oblivion. The buying trend toward four-wheel-drive vehicles has sustained the Jeep line, but the other AMC cars have not kept pace. A marriage of convenience with Renault was not the magic solution to its problems. Recently, Chrysler has purchased the brave and venerable firm, and ultimately will probably continue to produce only the Jeep Eagle.

LEFT:
General Motors, switching tactics almost year by year in a desperate attempt to stave off foreign and domestic inroads on what used to be an empire built on dogged owner loyalty, has been almost capricious with the Buick line's marketing strategy. Now, as GM quality and styling begins to make the long trip back, this Buick Riviera reaches for the conservative upscale image that had been Buick's heritage.
RIGHT:
The Ford Escort has been the best-selling nameplate in the world for several years. A classic case of real versus perceived value, it embodies reliable quality at an attractive price and is now dressed in the flowing sheet metal that has made Ford the envy of American automakers.

The story of styling has thus come almost full circle from the days of Benz and Daimler. Quality was paramount then, because maximum quality was necessary just to get the cars to run. A long period followed during which styling established its ascendancy over every other consideration (especially quality) and led to the production of millions of cars that could have been made much better, to everyone's benefit. Now we seem to be entering a period where styling has assumed a more appropriate role, as a useful adjunct to engineering expertise and manufacturing quality.

Perhaps the most encouraging indication of all is the fact that virtually all manufacturers have become increasingly dedicated to quality first. While Toyota and Honda reign supreme in the reports of consumer-protection magazines (just as they do in the road tests of the car magazines), less familiar names maintain a constant pressure on them to do well. Subaru, which (like Honda) once made only funny-looking little cars, now offers a full line of automobiles that command respect. The same is true of Mazda and Mitsubishi. Furthermore, Detroit has tried to bridge the gaps in its own production offerings by purchasing Japanese cars to sell under American name plates, as well as by buying foreign components for installation in ''native'' cars.

The most successful effort in this regard—and the one that clearly points the way to the future for every manufacturer—is the Ford Escort. It debuted in 1980 and was proudly touted by Ford as ''the

LEFT:
After gamely placing all its chips on making "Quality Job One," Ford let its bets ride with the radical-aerodynamic, low-drag styling on the Taurus and Mercury Sable models. It could have been the Chrysler Airflow-failure phenomenon all over, but the new Ford products caught on so swiftly that they have propelled the company to record-breaking profits.
RIGHT:
Probably one of the best cars in the world, with an impeccable record for reliability, the Toyota Camry is a consistent top seller.

World Car" because of its international origin. Designed and developed by Europeans, it was built on both sides of the Atlantic. The transmission and running gear came from Ford's Japanese affiliate, Toyo Kogyo. The Escort, continually improved over time, has been the best-selling car in the world for a number of years. If the Escort could maintain its relative price level and achieve the quality level of the Accord or Camry, the automotive millennium would surely be at hand. Perhaps the ever-growing competence of the industry with computers and robotics, together with a new sensitivity to consumer needs, will make such a goal realizable for most members of the world's car-making community.

AUTOMOTIVE ARMAGEDDON

A new threat to the stability of the automotive industry has appeared—one that has been discounted in both America and Japan, and yet may cause a domino-falling sequence of events with enormous consequences for the world's economy. There is currently a gap between the prices that must be paid to achieve the level of quality demanded in new cars and the quality available at a lesser price from used cars. The niche is being filled by various cars from previously unmentioned quarters of the globe. Korea has had phenomenal success with its Hyundai, which broke first-year sales records for a new import. The Hyundai's price amounts to about 80 percent of an equivalent American car and perhaps 60 percent of a comparable Japanese car (when the price premiums paid for the latter are considered). Its quality level, barely acceptable at its introduction, has improved considerably.

Other imports in the same arena—the Yugo, and the Dacia from Rumania—are even cheaper but are notably less successful, since the quality control, styling, and engineering are quite deficient. Yet there is no reason to believe that they will not improve. When (not if) they do, their effect combined with that of Hyundai and other Korean imports might well be to force an already stressed Japanese auto-making industry to cut prices on its cars just as the Japanese computer companies have cut prices on their microchips, dumping its products in such a deluge as to displace the American car makers entirely. It is not too far-fetched to look twenty years into the future and see America still producing lots of cars—all of them from Japanese plants located in this country. The danger of such prognostications is that they will be seen as calls for protectionism rather than as calls for improved productivity and quality at home.

OTHER TRENDS

In considering the art of economy being pursued in the present Relearning period, we have reviewed only the main thrusts. This period of time is linked to the previous era by the death and resurgence of the muscle cars—those bruising, powerful descendants of the California hot-rods. John DeLorean led off with the 1964 Pontiac GTO, which was essentially the marque's midsize Tempest crammed with the largest possible V-8 that would fit into it if you bulged the hood enough. Plymouth beep-beeped its way into the same market with the RoadRunner, Ford hissed in with its Cobras and Mach IIs, while Mercury's Cougar Eliminator snarled at the starting gate. It was all great fun, permitting you to go practically as fast as you dared straight ahead, even if you were foreclosed from taking turns at any great speed. The Chevrolet Corvette, born in 1953 as a fiberglass shell placed over standard bland Chevy components, has grown to be the classic American sports car. Equipped over the years with progressively more powerful engines and increasingly sleek body styles, the Corvette provides a relatively inexpensive challenge to Ferraris and Porsches in performance, if not in quality control.

In Europe, no upper limit is placed on the price of quality. Mercedes and BMW vie for top honors in Germany with cars ranging in price from £13,250/ $25,000 to £42,400/$80,000, and the two also compete with Jaguar in England. Italy markets the Maserati in smaller quantities at slightly lower prices. Rolls-Royce continues to sell all the cars it wishes to make at prices in the £53,000-/ $100,000-and-above range, while the Bentley is offered to buyers who are not confined to the Mercedes price range but feel that a Rolls is too much. The Bentley is not the only alternative in its class. The timeless Aston Martin Lagonda sedan is available, and you can always have a specialty firm such as Bitter or AMG refine your Mercedes—if that is not a contradiction in terms.

In the United States, there was still sufficient enthusiasm during much of the present period to support cars such as the underrated, Canadian-built Bricklin. A total of 2,897 Bricklins were built, and today they are rapidly becoming collectors' items. Part of their charm is that close to 1,000 of them were built at a factory cost of about $16,000 and yet sold to dealers for $5,400. An astute accountant sensed that something was wrong, but by then the company was $23 million in debt and had to shut down. A few years later, we enjoyed the DeLorean DMC-12. Built on faith and charm, some 4,000 were shipped from their Irish factory to the United States before the collapse of the DeLorean dream. Distinguished by its sleek lines and a stainless-steel body, the DeLorean is also going to be enjoyed primarily by collectors. Its advertising career was checkered, the low point occurring when a Wisconsin dealer offered ''60 pounds of Coke'' with each car, and showed a photo of a DeLorean stacked with Coca-Cola bottles.

A few very dedicated enthusiasts have created a covey of specialty cars. The first of these was Brooke Stevens's Excalibur, a 1965 Studebaker chassis and engine cloaked in a near replica of a 1928 Mercedes body, intended as a symbol of the Studebaker-Mercedes marketing connection. The Excalibur has grown in status over the years, changing its internals from time to time as the company has undergone various types of financial reorganization, but always maintaining a classic exterior.

Another spawn of Studebaker, the Avanti II, retains the original's styling with vastly improved performance and quality. Still other specialty cars —the Clenet, the Zimmer, the Stutz Blackhawk, and their ilk—offer distinctive styling and reliable drive trains to clients wealthy enough to have tired of Mercedes and Rolls-Royce cars and daring enough to want something different.

CHAPTER 5

NOBODY'S PERFECT: QUIRKS OF THE ROAD

Over the years, the unending stream of automobiles has been peppered like a hot Thai soup with great misadventures, miscalculations, misjudgments, and mischievousness. These departures from the generally staid automotive norm usually resulted from an excess of enthusiasm. The inventors—or perpetrators—were often men of great tunnel vision, who saw a problem or an opportunity and were determined to act. As a rule, they lacked the compensating insight to see that costs, efficiency, practicality, or other considerations worked against their aim; all too often, they were simply ahead of their time.

Normally such individuals see criticism as calumny, opposition as conspiracy, and failure as bitter vindication of a dream beyond ordinary human ken. But not every failure is steeped in eccentricity. Some of the most notable ones arose as great bureaucratic convulsions, as problems worked by many to conclusions that were absurd but could not be perceived as such because of the corporate effort invested. The Edsel and the Airflow are examples of well-intentioned, rational, totally inept major efforts.

And there is a third class entirely: ideas or cars that should have succeeded, if not for the intransigence of public vision or the ugly manipulations of competitive corporations. The Tucker is often thought to belong in this category—a car whose time had come but whose revolutionary characteristics were suppressed by the monster companies in Detroit (who in the process, of course, managed to hide the 50-mpg carburetor and the 100,000-mile tire).

Perhaps to make up for all of the heartache, there are soaring expressions of individuality in the cars created by individuals for themselves. Usually never intended for production (though a few, like Edsel Ford's Continental, found their way there), these wild chariots usually represent a personal dream car that speaks to the heart of its owner—if not to anyone else.

Actress Kathleen Burke with two 1935 "cars of the future," real quirks of the road.

Horses from Unusual Stables

Given the central importance of motive power to the entire concept of the automobile, it is not surprising that certain enterprising inventors were discontent with the prospect of being confined to gasoline, steam, or electric sources of power.

In 1895, a Mr. A. Burdick, of Hubbell, Nebraska, proposed a spring-wound automobile motor that would power a simple buckboard by means of a chain drive. What Burdick noticed (which Benz and others had missed) was that, while a gasoline engine could use only one fuel,

012·672
NY 35
19 CALIFORNIA 35
H 86 91
Y 11350
19 MICHIGAN 35

the spring-wound car could be wound by hand, by a small electric motor, or by rolling downhill. In a suitable area, it might be conceivable to start downhill, power up for the next slope, charge up on the downturn on stored power, and so on, practically indefinitely. Burdick wasn't the only inventor of a spring-wound vehicle, but he was every bit as successful as the rest of them.

A more serious effort toward a novel power train was developed by the well-established Woods Electric Company of Chicago, which built handsome electric cars for twenty years between 1899 and 1918. By 1916, the handwriting was clearly on the wall for electric cars, because of their dismayingly short range, so in response the company came up with the Woods Dual-Power, which retained the standard electric motor, but added a four-cylinder gasoline engine. The gasoline engine was used for starting, for operating at speeds up to about 15 mph, and for recharging the electric batteries. It is an idea whose time may yet come; but the model was discontinued by Woods within a year, and the firm was out of business within another year.

The influence of aviation has been pervasive, as in the rotary-powered Adams Farwell, but the 1925 Julian and the 1922 English North-Lucas went farther than usual by employing radial engines. Given the difficulty of cooling radials in aircraft, the task of doing so in an automobile was probably insuperable.

Water-cooled engines were different, so the Curtiss OX-5 engine (which had powered the famous World War I Jenny and most of the immediate postwar generation of sport planes) was called upon to serve as a powerplant for a number of cars. It had almost every conceivable disadvantage—great weight, difficult cooling, inappropriate running speeds—but it was available at low cost from government surplus stocks. Glenn Curtiss himself created a flying automobile, which reportedly was driven but never flown. Ben Gregory put a Curtiss OX-5 engine in a front-wheel-drive racing car, and this intrigued Curtiss sufficiently to lead him to adapt an engine to the well-known Phianna luxury car, whose factory was taken over by the Wright-Martin aircraft company in 1918. (The name *Phianna* was the combination of the names of two daughters, Phyllis and Anna, of one of the principals of the firm.) A number of other firms also saw merit in adapting the OX-5 engine to fit existing chassis. A 1910 Winton underwent the surgery, and afterward reportedly was able to top 100 mph—probably close to three times the Winton's previous upper limit. The Prado Motors Corporation of New York City is said to have built five or ten cars with the OX-5 engine.

In a more direct application of aviation technique, some manufacturers adopted the propeller as the source of motive power, shielding it with a circular duct, much as modern swamp buggies do. Curiously, these were almost uniformly successful in performing economically and at a good speed. The obvious hazard to life and limb, however, and the ear-splitting noise of the propeller were unendurable drawbacks—even for cars like the 1922 Reese Aero-Car, which could legitimately boast that it needed no transmission, belts, or gears.

The 1926 McLaughlin Maine-Mobile was a much more ambitious undertaking, with a 72-horsepower engine driving the vehicle as fast as 106 mph—or so McLaughlin claimed. That was on good roads, of course; it could also be driven at 35 mph (30 knots) in the ocean, and at an unspecified speed over the snow. The McLaughlin was not unattractive to look at, and it was rife with clever ideas, including the use of the car's wheels as rudders for its aquatic passages. Perhaps the first of all such endeavors was the 1904 MacDuff Aeropinion, which could top 16 mph on the strength of its 4-horsepower engine. With small wooden wedges affixed to its wheels, it became the Pneumoslito, or air sled—proving definitely MacDuff's mastery of coined verbiage as well as of coined automobiles.

In March 1950, the first public display of a turbine-powered car was made by the venerable English Rover company at Silverstone Airport in England. Called the Whizzard, the modified Rover 75 sedan reportedly achieved 93 mph and 5 mpg.

The next effort in this area was made in 1954 by Chrysler, which fitted an otherwise standard Plymouth Belvedere with a 100-horsepower turbine. Other experiments followed, culminating in the 1963 Chrysler turbine styled by Elwood Engle (who had designed the recent Ford Thunderbirds) and built by Ghia in Italy. Fifty were built, and an elaborate testing program was undertaken involving three-month "tours of duty" by volunteer temporary owners. It was a stunning-looking car and extremely responsive, capable of going from 0 to 60 mph in less than 10 seconds. Not unnaturally, it was also fuel-hungry and got only about 12 mpg.

The public loved them, but Chrysler was forced to destroy forty members of the series (to avoid paying import duties) and placed the rest in museums around the country. As with the aborted SST, the decision to withdraw was probably a sound one, given that the fuel shortage was less than a decade away, and it would have been extremely difficult to satisfy the government's CAFE (Corporate Average Fuel Economy) requirements with a line of turbine cars.

Milton O. Reeves got started in motoring in 1896, spending the money generated by his successful pulley business on a series of ill-fated automotive innovations (e.g., a muffler, a simulated horse's head on the grille). His greatest stroke of genius debuted at the 1911 Indianapolis Speedway Race, the eight-wheeled "Octoauto." Sadly, there were no buyers for either the "Octoauto" or its six-wheel follow-on, the "Sextoauto."

But the ultimate word in power (yet to be realized) was the Hungerford Rocket, built in Elmira, New York, in 1929. Shaped like one of the vehicles from the Buck Rogers serials, the Hungerford Rocket was capable of traveling 70 mph on rocket power. An internal combustion engine enabled the car to handle normal road work, as well. Brothers Daniel and Floyd Hungerford were quite serious about their effort and only wished that they could either have manufactured the Rocket in quantity or (at least) have built a bigger one with four rocket motors. Considering today's emissions-control problems, it is perhaps just as well that the Rocket remained an experiment.

Are Four Wheels Necessary?

As fascinating as alternative engines were to the backyard craftsman, even more interest was

R. Buckminster Fuller's 1933 Dymaxion was one of his best—and least-realized—ideas. A three-wheeler, powered by a Ford V-8 engine, it had a then sensational top speed of 120 mph *and* a 40-mpg fuel economy (both figures roughly double that of the average car of the time). Of the three that were built, including one for Leopold Stokowski, only one remains—in the Harrah collection in Reno.

OPPOSITE:
The Jaray car had its genesis in the wind tunnels in which Germany's World War I Zeppelins were designed. Its emphasis on achieving high performance through low-drag design simply came fifty years too soon.

piqued by the possibility of creating other forms of chassis, transmissions, and wheel arrangements.

In the configuration of wheels, three and six were the favored alternatives to four, although attempts were made to use one-, two-, and eight-wheeled arrangements.

Six-wheeled cars appeared first in 1903, with the French Janvier and the American Pullman. The name of the latter was a simple commercial theft of the then-prestigious Pullman corporation. A two-cylinder engine drove the middle pair of wheels, while the other four wheels were used for steering. It was not successful because the middle wheels had a tendency to lose traction. The French car was far more conventional, with the (four) larger rear wheels transmitting the power from the engine to the road, while the two front wheels turned to steer.

A flurry of six-wheel cars appeared in the next few years, all without much success. In America, however, the cars of Milton Reeves achieved lasting fame in the pages of Floyd Clymer's nostalgic books on automotive history. Reeves first created the 1911 Octo-Auto, an eight-wheeler that rode too well on smooth roads and was dangerously difficult to handle on rough ones. Then he tried again with a Sexto-Auto, before giving up.

But before dismissing the six-wheel idea entirely, we should consider how many trucks use the very same layout to help them carry their loads. Similar consideration prompted the building of the 1932 Hudson six-wheelers for the use of the Japanese army in Manchuria, while Mercedes-Benz adopted the configuration for its handsome Type 500. Even some racers used the scheme, including a Pat Clancy Special that ran at Indianapolis.

Three-wheel cars have had more success than their more multiwheeled brethren. The English Morgan established itself as a cult car on the performance of its three-wheeler, which was built until 1952. But of the cars that didn't enter series production, the most famous of all are R. Buckminster Fuller's Dymaxions. Based on the publicity they received, you might assume that Dymaxions were made in the thousands. Actually there were only three, but their performance and their promise render the publicity they received well-earned.

Fuller called almost everything he did *Dymaxion,* including his widely used geodesic dome construction. His cars, built in 1933 and powered by a Ford V-8 engine, were capable of 120 mph and more than 40 mpg—incredible performance characteristics even by today's standards. Large, comfortable, and capable of turning in a radius defined by its own length, the Dymaxion deserved a better fate. But sent on tour, it suffered a collision with another car, and its occupants were killed. The press seized upon the accident as ''proof'' of the car's instability, and the adverse publicity hastened the end of the project. Perhaps fittingly, the only one to go into private hands was sold to Leopold Stokowski, the bold, imaginative conductor of the Philadelphia symphony orchestra.

The Dymaxion inspired a number of other inventors to follow the same basic path. The 1936 Tucker (no relation to the postwar car of the same name) virtually duplicated, in miniature, the Dymaxion's layout and adapted aircraft steel-tubing structure to its rear-engine layout. The 1936 Aerocoupe and 1937 Airomobile were even smaller versions, pursued with less success. Perhaps the only claim to fame of the rather handsome Airomobile was

C.
+392

that its engine was manufactured by Doman-Marks, a company that was eventually purchased by Preston Tucker as he was setting up his company to build the postwar Tucker automobile.

The Dymaxion inspired some four-wheeled descendants, as well. The first of these was the Scarab, built by William Stout—the man who had induced Henry Ford to invest in airplanes and who inspired (if he did not design) the Ford Tri-motor. Stout built five Scarabs, all Ford-powered, featuring a versatile interior that would be familiar to van owners today. A more successful car, designed along the same lines, was the model built for the McQuay-Norris Company in St. Louis. These six streamlined rolling laboratories were powered by Ford and laden with test equipment appropriate to the company's interests. Never intended for production, the cars didn't receive the attention they deserved from the press.

The postwar years saw a host of three-wheel cars burst upon the scene, the two most interesting of which were the English Bond Minicar and the American Davis. The Bond was a tiny two-seater, entirely disconcerting to encounter on an English motorway because it moved at unseemly speed and possessed unusual agility. How it would fare in an accident is another matter, but the English loved them through the years. The Bond was transformed over time from a most austere and comfortless automobile into the luxurious and refined Reliant.

The 1947 Davis was intended to be totally different from most three-wheelers, offering itself not

Rust Heinz designed a special body for a Cord 810 chassis powered by an up-rated Lycoming engine, and called it the "Phantom Corsair." When used in the 1938 film *The Young in Heart*, the name was changed, Hollywood fashion, to the "Flying Wombat." A limited production run was planned, but never realized. It is seen here with Janet Gaynor and a companion.

as an economical substitute for "normal" automobiles but instead as a high-performance (116 mph), extraordinarily maneuverable new wave in transportation, to be sold for about $1,000. The whole project was terminated by a messy investigation into the company's financing. The question of fraud was raised, but there was no question about the capability of the vehicle, which was demonstrated in hair-raising maneuvers performed at high speed and with total safety.

TV trivia buffs will remember that Chester Morris drove a Davis in his television series "Johnny Dollar." It was the perfect car for a cop or a crook, because the single front wheel could be turned 45° in any direction, and the car could be driven at close to 60 mph in a circle smaller than the turning radius of most cars of the time. Imagine the chase scene that could have been created: a Lufberry circle of cop and crook circling at 60 mph in an intersection! The Davis featured a single bench seat for four, with a detachable metal top; seven- and eleven-passenger versions were also planned.

Sonia Delaunay designed her clothes and the finish of the Citroën B12 in 1925, anticipating Calder's work for Braniff Airlines by forty years.

With the matter of how many wheels to use being pretty well settled by time and custom at four, another issue to be decided is which wheels should be powered. The traditional Panhard configuration ruled until the advent of the Morris Mini-Minor and the subsequent wholesale adoption of front-wheel drive by the rest of the world. Prior to this time, the term *front-wheel drive* conjured up (in America) the Cord, with all its problems, and (in France) the fabled Citroën Traction Avant, the car preferred by gangsters and Jean Gabin.

As long ago as 1897, a clever blacksmith named Joseph Barsaleaux invented a five-wheel, front-wheel-drive horseless carriage that also had a horse. The horse was a dummy, of course, intended to comfort the real horses on the road; but the vehicle was actually steered by reins running to the conventional bit, comforting drivers who were trained in driving a horse but not a car. With such a beginning, it is not surprising that many front-wheel-drive cars appeared early in automotive history—many of them quite capable. It must be admitted that most of them adopted the layout as a styling tool and that most suffered from grave reliability problems. But the basic advantages in space saving, traction, and weight distribution were worth the design effort.

J. Walter Christie, a talented, largely self-taught engineer, was the first of the front-wheel-drive enthusiasts, and he established patents around the world for his version of the device. A gifted inventor whose ring-turret turning lathe was used as the model for the gun turrets installed on the second *U.S.S. Maine,* Christie built front-wheel-drive racing cars that competed in the French grand prix. His cars featured huge engines (a V-4 of just under 20,000 cc!), and he spent a lifetime designing tanks, gun carriages, amphibious vehicles, and airborne tanks—all of which he offered to the United States government, and none of which were adopted here for quantity production. His tank and gun carriage designs were used extensively in both England and Russia, however.

The Depression called forth a series of front-wheel-drive cars, all handsome and all expensive. The famous Cord L-29 is the best known (and perhaps the best car), but the 1930 Hamlin and its brother beneath the sheet metal, the Hoffman, were attractive and featured a unique twin-axle front drive. The weight of the car was suspended on one axle, while the power was transmitted through another. Harry Miller, whose front-wheel-drive racers achieved enormous acclaim, built a few custom passenger cars (for as much as $35,000 each), but the owners didn't seem to care for them.

Perhaps the most intriguing of all the Depression-era front-wheel-drive cars was the Ruxton, a car that brought the kiss of death to every company involved with it. It was not the fault of the car, but of the times and of the incredibly destructive management tactics of a man named Archie Andrews, who seemed to be present at the demise of almost every car company that went under in the 1930s. He was involved in a series of incidents in which his control over (or attempts to control) a manufacturer resulted in the company's dissolution. Andrews's intentions were undoubtedly the best: he wanted to introduce a new era in cars, announced by the front-wheel-drive Ruxton. He had no factory, however, and he went through a series of savage business battles to get what he wanted from those who did. The final one actually saw the principal officers of the venerable Moon Company lock themselves into the board room, only to have the doors battered down by Andrews and his lawyers.

Powered by a 100-horsepower Continental engine with a unique divided gearbox, the Ruxton was only 63 inches high at a time when most cars topped the 6-foot mark. Its low-slung styling was accentuated in some models by gorgeous eight-color horizontal striping; and its sexy looks were emphasized by its cat's-eye Woodlites and by the absence of running boards.

In the end, the Ruxton's protracted delivery schedules and general financial difficulties coincided with the general decline of the luxury car during the Depression, and only about 400 to 500 of the cars were ever built. They nonetheless had an excellent reputation. Not as handsome as the Cord L-29, and priced $100 higher, the Ruxton at least shared the Cord's Moon ancestry.

But as has been indicated many times heretofore, no matter what the engine, no matter what the transmission, the thing that really sells automobiles is the styling. And in styling, there were quirks aplenty through the years.

STYLING: BEYOND THE TRENDS

The Ruxton and the Cord L-29 inspired a *New Yorker* cartoon in which a very Ruxtonish sedan was pictured, with the caption "At Last! A Car So Low, No Human Can Get In It!" Indeed, over the years many of us have gladly suffered blows to the head and knee for the pleasure of riding in low-slung, streamlined cars. And if the ride is uncomfortable, what of it? Appearance is everything. People who drive Lamborghinis and Ferraris despite their lack of comfort must do so for the simple reason that heads turn when they go by.

Streamlining seems to have elicited the best and the worst from designers. For some, streamlining was the sole object of styling, no matter how it was obtained, and no matter how impractical or ugly the result was. For others, what was sought was verisimilitudinous streamlining—the appearance, but not necessarily the reality of aerodynamic perfection.

The design process was a learning process. Netherlands-born John Tjaarda had cut his teeth in a number of design studios, including Harley Earl's Art and Colour Department, before joining Briggs in 1932. He had also created the audacious Sterkenberg series of rear-engined beetle-shaped streamlined cars but had been unable to find financing for them. At Briggs, his design genius reached maturity and resulted in front- and rear-engined prototype cars that greatly influenced the design of the Lincoln Zephyr.

The 1935 Tatra 77 had a 70-horsepower, V-8, air-cooled engine that gave a top speed of 90 mph. A monocoque body was mated to a tubular frame, providing a very roomy interior.

Most other efforts were not so felicitous. Even the immortal Bugatti was not immune from design errors in attempting to streamline cars, once making a series of cars in the shape of half-walnuts, oval from front to rear but perfectly flat on the sides. The equipment and technique necessary to mold complex compound curves were not yet available to him, and he was willing to accept the simplistic, ugly shape as the best attainable form of streamlining.

If Bugatti could err, then why not Norman Bel Geddes, who designed more cars, trucks, and buses that were never built than almost anyone else in history? Apparently a lovable individual who could talk his way into any contract, Geddes is most famous for the work he did on the General Motors Futurama exhibit at the 1939 New York World's Fair. In it, thousands of unlikely miniature Geddes-designed model cars scooted around amongst futuristic scenery. His last automotive experience was with Nash, where his models may have influenced the (perhaps) unloveliest production car ever built: the Nash Airflyte series of 1949. Almost always attributed to Nash's chief engineer Nils Wahlberg, the cars were quintessentially Geddesian in appearance—bulbous of line, with wheels tucked away like four tiny bound feet. These cars very likely started Nash on its way out as a full-line manufacturer.

But the dean of streamline design was the Aus-

The 1937 Airomobile.

trian Paul Jaray. His training in aerodynamics led to a position designing first aircraft and then Zeppelins in Count Ferdinand von Zeppelin's Friedrichshafen works. There he built a wind tunnel for testing both lighter- and heavier-than-air designs. During the process, he worked in concert—and to some extent in competition—with Claudius Dornier. The best of Jaray's aircraft designs was the two-seat observation plane adopted by the Swiss Air Force after the war.

The wind-tunnel experience he gained working on Zeppelins and planes convinced Jaray that his ideas on streamlining could be applied to automobiles. He conducted a series of scientific tests comparing the standard (almost square) cars of the time with cars made according to his own sleek design. Jaray saw the ideal car as having a smoothly rounded body, a fully enclosed glass cabin, and a duckbill-like rear. Its drag coefficience was a near-miraculous .245—a figure that would win headlines for any modern production car.

Jaray continued experimenting, and his ideas were sampled by dozens of manufacturers. Only one, Tatra of Czechoslovakia, built his cars in series, but he negotiated contracts by which some gorgeous prototypes were built by Maybach, Audi, Mercedes, Hanomag, Opel, and Fiat. Other firms simply infringed on his patents, including Chrysler, Willys-Overland, and Pierce-Arrow. Sadly, Jaray never received the financial compensation for these borrowings that was clearly his due.

Into the Sea and into the Air!

Most of us can remember the German Amphicar, imported into the United States between 1961 and 1968, which has recently begun to receive wide attention among collectors. Priced at only $3,395, it was the only amphibious production passenger car in the world, drawing on thirty years of military amphibious vehicle design. A Triumph Herald four-cylinder engine propelled it at speeds of up to 68 mph on the road, and 7 knots (roughly 8 mph) on water. The front wheels performed double service as rudders. The car was fairly innocuous-looking and had an unfortunate (especially for an amphibious vehicle) tendency to rust out. The real problem was its price: a Chevrolet Impala sedan could be purchased for about $800 less than the Amphicar—enough money to get a start on a boat and trailer, if water transportation were desired.

Much earlier in history, a very attractive-looking vehicle, the Hydromotor, was built by the Automobile Boat Manufacturing Company of Seattle. Steered in the same direction whether ashore or afloat, the Hydromotor claimed 60 mph on land and 25 mph (21 knots) on water. The car was tested in 1915 and was being readied for production in 1917, but there is no record that any car other than the prototype of the Hydromotor was ever made.

In 1917 George Monnot designed the Hydrocar in Canton, Ohio, powering it with a dependable Hercules four-cylinder engine. It differed from most vehicles not only in its amphibious nature, but in its being designed to be driven in one direction on land and in the reverse direction on water. Top speed on land was 25 mph, and top speed on water was about 8 mph (7 knots). It was tested by the military, and—had the war gone on—it might well have been placed in production.

The best-looking and most promising of the car-boats appeared in 1940, when Paul Pankotan created a slick-looking cruiser fitted with four wheels that were mounted in such a way that they could be raised for water travel. A 90-horsepower gasoline engine was supposed to be capable of delivering 90 mph on land and 35 mph (30 knots) on water—probably an optimistic estimate, but still not too unreasonable. No one picked up on the idea then; but with the modern, nonrusting materials now available, its time may have come.

With the exception of the Curtiss combination car and airplane of 1917, there was little pre–World War II activity in the car-plane field. Various combinations of automobile engines were fitted to airplanes—the Ford V-8–powered McGaffey Aviate, Fahlin Plymocoupe, and Arrow—all promising economy, but all failing to deliver the required

PAGES 196–197:
The 1937 Airomobile was another economy car ahead of its time. It was intended to be a very inexpensive car with an air-cooled engine, front-wheel drive, and only three wheels. No one felt like financing the car, however, and it succumbed to the Depression.

performance. Automobile engines have been adapted to aircraft use more successfully in recent years. Air-cooled Volkswagen and Porsche engines have found widespread experimental use, and there has been a recent trend toward adapting Jaguar, Buick, and Oldsmobile powerplants to both home-built designs and aircraft proposed for production.

Historically, there have been a few true airplane-car combinations. Waldo Waterman, one of the great pioneers of aviation's byways, built a series of Arrowbiles (sometimes spelled *Aerobile*), featuring distinctive layouts and easy disassembly of the wings for road use. The Studebaker company became infatuated with the idea in the mid-1930s and seriously contemplated selling the Arrowbile through its dealer network. Six-cylinder Studebaker Commander engines were to have served as the powerplant. At least one Arrowbile survives, powered with a more conventional (but still automotively derived) Franklin horizontally opposed six-cylinder engine.

The Fulton Airphibian was demonstrated in the years between 1946 and 1952. It was essentially

The big Duesenbergs ran so fast that it often proved to be an irresistible temptation to take off a few pounds of fender and bumper and race them. Here, the stripped-down car's escort is Gary Cooper.

a small two-passenger car that could accommodate a wing, propeller, and additional afterfuselage. As is inherent in most such designs, the package was a compromise that didn't perform very well in the air or on the ground.

A later attempt at a similar idea had a tragic conclusion. The Aircar had a Ford Pinto auto component attached to Cessna Skymaster wings and empennage. Unfortunately, it suffered a structural failure in the air and crashed in 1973, ending development.

The premier airplane-car designer of all time is Molt Taylor of Longview, Washington, who has designed many high-performance personal aircraft; his Aerocar came the closest to being a genuinely useful combination vehicle. Production was planned at one time in a Fort Worth factory, but the usual financing problems supervened.

The desire to build a combined car-boat or airplane-car can be spurred by native mechanical ingenuity or a need for self-expression. In the latter category of motivation, many noteworthy contributions have been made to our culture in the form of ordinary road cars carefully suited to the owner's psyche.

The Very Personal Car

The pop-psychological aphorism that a man is only a boy with more expensive toys is perhaps never more true than when applied to affluent car lovers. Clark Gable or Gary Cooper may have competed with custom-body Duesenbergs, and they may have liked cars, but they didn't *love* cars. The difference is not necessarily a matter of total cost, for an almost limitless amount of money could be lavished on special bodies and chassis (including gold plating). What is required to merit inclusion in the car-lover class is the need for artistic automotive self-expression, and this in turn requires a reasonable knowledge of available power trains and body designs.

Sometimes a single intoxicating foray into such self-indulgences starts a dynasty of similar efforts, as happened in the Fageol family. Frank and

ABOVE AND OPPOSITE:
In 1947, thousands of people eagerly adhered to the slogan "It Will Pay You to Wait for the Tucker 48." What a bombshell this car was on the automotive scene! Although only a few were built, they inspired a cult loyalty that is bound to be enhanced by Francis Ford Coppola's new film *Tucker*.

BELOW:
The 1965 Mercer-Cobra was designed by Virgil Exner to show how the automobile uses copper and its alloys: The Cobra was trimmed in eleven different hues of copper, bronze, and brass.

William Fageol began by building a $19,000 luxury production car in 1917. The war interrupted their scheme after only a few, powered by 125-horsepower Hall-Scott airplane engines, had been produced. But the fever had taken hold, and merely waited to be indulged later when the family fortune had been made producing trucks and buses.

Louis Fageol, a son of one of the founders, built a Fageol Supersonic in 1948. Low, sleek, and equipped with a slide-away sun roof, the Supersonic was powered by a 275-horsepower engine that used propane gas as a fuel (quite a radical innovation for the time). It was extremely practical, besides being head-turning, and the Fageols used it regularly as a family car. While not truly supersonic, it did top 135 mph.

Louis's son, Raymond, followed in the family tradition with his unique Fageol Special—a long, low-slung car that looked like an Indy racer with an envelope body. Less practical than Louis Fageol's Supersonic design, the Special was nonetheless clearly in advance of industry styling, with its flow-through fenders and outsized fin.

Normally, manufacturing difficulties kept the amateur-designed personal car from attaining genuine beauty. But if the Chrysler Corporation had consulted Vincent Bendix on the styling of their abortive Airflow line, a disaster might have been turned into a success. In 1934, Bendix—the automotive parts manufacturer who sponsored the Bendix Trophy for the National Air Races—created a handsome personal four-door sedan he used for a number of years. The car used all the applicable Bendix products, including a vacuum shift, brakes, and a special front-wheel drive. There were hints of the Pierce Silver Arrow in the body work, but the overall execution was superior, and Bendix's personal car was probably the best American streamliner of the time.

The most famous streamliner, however, was undoubtedly the Phantom Corsair, built in 1938 by Rusty Heinz, an heir to the Heinz 57 fortune. A Cord 810 chassis formed the basis for Heinz's coachwork, which was executed by Maurice

Tucker

Schwartz of Pasadena. The Phantom Corsair cost almost $25,000 to build, but Heinz intended to market it in limited numbers. Unfortunately, he died in 1939, and the car achieved its greatest fame when it came into the possession of entertainer Herb Shriner. It can still be seen on late-night TV occasionally in a movie entitled *The Young in Heart,* where it is called the *Flying Wombat.*

The latter name does fit the car better than *Phantom Corsair,* for the design had many more faults than virtues. A full-envelope body and fully skirted front wheels imparted a tubby look that was accentuated by the small glass area. The car lacked balance and distinction; and its notoriety is attributable to its being different, not better.

To see a car that was both different and better we need only turn to Preston Tucker's postwar classic, which arrived on the scene with an emotional charge that would be impossible to duplicate today. Billboards and posters said, ''It will pay you to wait for a Tucker 48,'' and the automobile-starved masses of America believed this wholeheartedly. For while all of the major manufacturers were serving up 1942 models with minor cosmetic changes, Preston Tucker was offering the car of the future. It came in an external package that remains eyestopping to this day. Designed by Alex Tremulis, the Tucker had long, flowing lines, doors that reached into the roof (the design features of which were duplicated by Chevrolet for the Corvette, years later), and three headlights—the center one moving with the wheels!

Is it any wonder that crowds thronged the special stock-selling auto shows promoted by the corporation, or lined up to put delivery deposits down and then lined up again to buy luggage custom-made to fit into the Tucker trunk they already had ample reason to believe they would never own? There was probably never a greater degree of would-be-owner loyalty than that among prospective Tucker owners. Even after the bloom had begun to fade from the rose of Tucker's financing, people invested because they *wanted* a Tucker.

Preston Tucker had been born and bred in the automotive business, working with Harry Miller and others in the hands-on end of automotive engineering. But Tucker—not unlike E. L. Cord—was a master salesman, and his greatest selling job came when he acquired (while possessing virtually no assets of his own) an enormous 475-acre plant from the War Surplus Administration. Located in Cicero, the Riviera of Illinois, the plant had formerly been operated by Dodge to build engines for the B-29.

Tucker's concept for the car was revolutionary. The air-cooled engine mounted in the rear was designed to drive torque converters to provide a smooth power train, low profile, and good weight distribution. The passenger compartment featured a cavernous ''bomb shelter'' into which passengers could leap in case of an accident, and windshields that popped out on impact.

In a single week, Tremulis put all of Tucker's ideas together into a beautiful package. Within 100 days, the famous Tin Goose prototype Tucker was built. Like all such rush jobs, it was suitable for auto shows but not for the road, being filled with lead, and naturally it became the object of rumors that it wouldn't run, had no reverse gear, and so on.

Fifty other cars were built in the Tucker factory, on a semiproduction line. These were essentially pilot models, intended to qualify the company for inclusion in *Ward's Automotive Journal,* to spur dealers to buy franchises and the public to buy stock. Inevitably they had compromises of the sort that new designs must always tolerate.

But even with the compromises, material and parts shortages, a skeptical press, and hostile government investigators, the Tucker that was produced was a remarkable car. An air-cooled Franklin engine was converted to liquid cooling and provided 166 horsepower that echoed with an unforgettable resonance through six exhaust pipes. Some were fitted with Cord preselector transmissions. The unimpeachable Tom McCahill said that the Tucker was the best-performing car in America by far, based on its 0-to-60 time of 10 seconds and its top speed of over 120.

But the Securities and Exchange Commission found fault with the sale of $15 million in stock,

and the ensuing litigation caused the whole enterprise to collapse. Tucker, ill with cancer, fought back as best he could, but he died a bitter and unhappy man.

Tucker fanciers were bitter and unhappy, too, because the Tucker was a car worth waiting for. The extent of the American driving public's emotional involvement can be measured by a comparison of the Tucker experience with the Kaiser-Frazer experience. Tucker failed with only $15 million, while Henry Kaiser failed with more than three times that. A gallant remark is attributed to Kaiser to the effect that he didn't mind throwing $50 million into the automotive pond—but he hated to see it disappear without a ripple.

Far better financed, and with legislative and business clout that greatly surpassed Tucker's, Henry Kaiser should have succeeded in the car business. Equipped with a larger plant than Tucker's—the enormous Willow Run facility (inspiration for a predictable outpouring of "Willit Run" jokes)—Kaiser joined forces with Joe Frazer, who ran the remains of the Graham-Paige organization.

The initial series of Kaisers produced in 1947 sold very well. Even though they cost more than the comparable Big Three makes, they were good cars with competitive performance; and best of all, they were available for delivery. But as America geared up and the sellers' market disappeared, the bell began to toll for the new manufacturer.

The vision of California designer Phil Garner—combined with what he saw as the marine "inspiration" of the 1968 Buick Le Sabre—led to the creation of this Nauti-mobile. The addition of a flying bridge makes it possible to skipper the car through even the most treacherous traffic conditions.

Holding valiantly to the clapper was Howard Darrin, who created the best-looking car of the period—the breathtaking 1951 Kaiser, with its sweeping lines, expansive glass area, and distinctive trim. But it was still overpriced, and (perhaps fatally weakening it) it lacked the V-8 engine necessary to match the performance offered by such competitively priced cars as the Rocket 88 Oldsmobile. After a run of perhaps 700,000 cars, another valiant experiment in independent car production came to an end.

The passage of time and the current market value of nostalgia has generated a high degree of interest among collectors in quirks of the road. This interest has been augmented by the visual and emotional impact of designs from the great era of classic styling, so that car restoration is today a multimillion-dollar business that has spawned dozens of minor adjunct industries. Best of all, it has brought thousands of cars of every make back from the grave and into the spotlight. Automobiles that might have been bought for scrap (and reduced to folded metal cubes) thirty years ago are now selling in the seven-figure range.

C H A P T E R 6

RE-CREATING THE PAST

Parades of Phoenixes

In April 1987, at the Imperial Palace in Las Vegas, a great deal of money changed hands, and none of it a gamble. A glittering collection of cars, almost every one perfectly restored, and others so nearly unused as to pass for new, were being offered in a no-reserve auction. The selection ranged from a 1902 Haynes Apperson Tourer (it was a Haynes Apperson that raced from Kokomo, Indiana, to New York City in a sizzling seventy-three hours) to a Type 57 Bugatti Ventoux coupe to any number of Rolls-Royces, Bentleys, Cords, and Duesenbergs.

Unlike most events in Las Vegas, this was one where nobody could lose. Purchasers who bid successfully, at no matter what price, need only hold onto the cars to make a profit; while cars that did not sell will quietly continue to appreciate for their owners.

It was not always this way. During the Depression, luxury cars depreciated at incredible rates. A Pierce-Arrow purchased new in 1931 for $4,000 might be found on a used car lot three years later with 10,000 miles on the odometer, a $500 price tag, and no takers. Gasoline was terribly expensive, of course—perhaps as much as 15 cents per gallon—and 1 gallon of gas would take a Pierce only about 12 miles. Smaller cars (Fords and Chevrolets, in particular) held their value better, but after five years or 50,000 miles, they sold for small change.

Part of the problem was that the American psychology was attuned (after long exposure to reinforcing advertisements) to the desirability of a new car every year, and part of the problem was that conditions were so bad that few people could afford any car—new or used—at any price.

Other factors contributed to the wasting of restorable cars. The concept of collision insurance (and even of ordinary liability insurance) was foreign to most people. Consequently, if a car incurred some damage, it was often deemed too expensive to repair and was perforce consigned to a junkyard.

But as World War II approached, the need for scrap metal for munitions drained these yards steadily; and the biggest cars were always the first to go for this purpose, because less demand existed for their components. Then, when the United States entered the war, scrap drives began in earnest, and junkyards ruthlessly pruned their collections for all but the most contemporary material needed to supply parts for the few cars remaining on the road.

The absolute indifference to the value of older

Second-hand tires displayed for sale; San Marcos, Texas—August, 1940. *(Photograph by Russell Lee)*

cars continued until the late 1940s, when a mild interest first came to be asserted in the obvious classics—big Packards, Cords, Duesenbergs, and the like. But it was a light-hearted thing: someone might buy a big Cadillac in excellent shape, and drive it for a few years without any thought of restoring it to as-new condition. Such purchases were more an idiosyncratic expression of individuality than an intentional step toward the creation of a cult that worshipped restored cars.

The motor magazines were instrumental in changing this. The desperate need for story material (which all magazines experience) resulted in a demand for nostalgia articles. These were provided by some extremely knowledgeable people such as Ken Purdy, Griffith Borgeson, and Eugene Jaderquist, whose books and articles gave legitimacy and purpose to the activities of collection and restoration.

Soon, major collections began to be assembled. William Harrah of Reno, who began in 1948 with two cars, had thousands by the time of his death in 1978. Harrah had the funds to restore cars to as-new (and often, better-than-new) condition, and he set the popular standard of what restoration meant.

As the collecting fever spread, at least three things happened. The first was the inevitable rise in prices, which have in recent years approached those of fine art masterpieces. The second was a widening of the interest to embrace every style and make of car. The third was a broadening of the search, drawing thousands of cars from barns, garages, basements, attics, storage garages, and even open fields, to be transformed from junk into something approximating gold.

Perhaps a fourth thing should be mentioned. The cars found in barns and basements ceased to be bargains as owners became aware of their worth. The times when an M.G. TC might be given away no longer exist.

A mystique grew up surrounding the Concours d'Élégance meets held at fashionable spots around the country. Car-judging standards were established at levels that almost certainly would have excluded from consideration any car in merely factory-new condition.

To support the restoration efforts, all of the familiar impedimenta of a major hobby began to grow up. Organizations were formed, the Auburn-Cord-Duesenberg Club being among the first and most active. Newsletters and magazines sprang into existence. Swap meets began to be held, and it was found that, not only were there cars waiting to be restored, there were also huge supplies of parts for many of them.

For others there were not, and small companies appeared, dedicated to filling the urgent need for Rickenbacker clutches, Lincoln pistons, Cord transmissions, and Studebaker upholstery. In a typical recent issue of *Old Cars Weekly*—a marvelous specialty newspaper published in Iola, Wisconsin —advertisements offer new Pierce-Arrow spring shackle sets, V-16 Cadillac head gaskets, weather-stripping and running board mats for Terraplanes, motor mounts for 1932 Chryslers, and vent windows for 1934 Buick convertibles. If a part is not in stock, it can be remanufactured exactly to original factory specifications.

If you prefer not to do your own work, you can turn to any of hundreds of restoration shops that

How many car lovers have fantasized about finding the Cord or Hudson or Model T of their dreams languishing in some backlot, just waiting to be liberated and brought back to life? *(Photograph by Bohumil "Bob" Krčil)*

will be glad to do it for you, for a price. Generally these are specialists that confine themselves to one or two marques. And once you have a car restored, you'll want to avoid driving it (after all, what if someone ran into it?), so you can choose from among dozens of trailer manufacturers eager to provide exactly the custom trailer you need to haul one to four of your antiques.

Varying degrees of hauteur can be found at different contests and meets. At the annual Pebble Beach Concours d'Élégance, the most beautiful cars in the world are trailered in. Each car is immaculate, with no fleck of dust to sully the bottomless layers of flawless paint and not a spot of grease under the hood or upon the chassis. The show usually features a theme marque, so that all of the existing Bugatti Royales, or the rarest and most exotic Mercedes, or a bewildering variety of Duesenbergs will be assembled in one gorgeous location, amongst any number of gorgeous people.

There are even a few races in which vintage cars compete against each other at speed. In general, winning is less the aim than demonstrating how well the car handles. Once again, the most prestigious of these events is in California, at the Monterey Historic Automobile Races. In 1986 the honored marque was (fittingly enough) Mercedes-Benz, celebrating its and motordom's 100th anniversary. The races were humanized both by the spirited demonstrations of superb machinery and by the gathering together of many legendary racing names of the past. Juan Michael Fangio, Herman Lang, Phil Hill, Jackie Stewart, Stirling Moss, and Briggs Cunningham gave impressive demonstrations of the driving skills that earned them enviable fame, fortune, and longevity.

Still other clubs sponsor tours, in which the cars are driven by their owners on an intercity journey. The owners dress in contemporary costumes, the cars break down and must be repaired, and a good time is had by all. Each run is a microcosm of times past—for the cars, meticulously restored and painstakingly serviced though they are, are still assemblages of old parts.

On the way to the auto show—a popular, if expensive, contemporary hobby. Anyone with the foresight and skill to restore a 1933 Auburn Phaeton to this condition is probably going to make more money this way than if he had invested equivalent funds in the stock market, however. ***(Photograph by Curtice Taylor)***

Restoration: Putting Humpty-Dumpty Together Again

Flirting with old cars is dangerous, because a time may come when the desire to own one changes from a fancy to an obsession. Thus, it might be instructive to review the restoration process, step by step, from the time the automobile arrives in the garage—displacing all of the lawn furniture, winter clothes, and other intended contents of garages—to the time when it is ready for its own special house because the garage just isn't good enough.

Photographs and data are the first order of the day. The car is photographed inside and out, and each part, as it is removed, is photographed, as well. This accomplishes two things: the car's present state is recorded for posterity; and a sequence is established for reassembly. Simultaneously, efforts are made to obtain period handbooks on repairing the particular type of car at hand. Contemporary works on the general process of restoration are also consulted.

One of the first hazards involves determining what is and what is not original to the car. Everything from nonstandard engines to nonstandard bodies has been discovered at this stage. As the parts come off the car, two inventories are maintained: one inventory of parts on hand, and one inventory of parts needed. Ordinarily, the latter list is longer than the former.

In the process of disassembly, difficulties arise due to the natural effect of time on lubricants, the deterioration of the many organic materials (cloth, leather, and the like) that were used in vintage cars, and, of course, corrosion. Rust is the enemy, and it strikes at every level, from the obvious inroads made on body panels to subtler (and costlier) destruction of radiators, engine blocks, brakes, chassis, fuel tanks, and anything else made of metal. In a similar way, rot attacks wood, which was an essential component of bodies until the late 1930s.

As the inventory lists grow, their most striking feature is their enumeration of so many specialty items that cars were and are composed of and that people are normally unaware of. An engine is a comprehensible chunk, and a water pump or a distributor, too, might be anticipated to be missing and to require replacement. But the horseshoe retaining clips for the clutch, the leather gaskets for the carburetor, the special screws for holding the dashboard fascia, the glass cover of the ceiling lamp, the brass strips for establishing electrical connection for the bulb, the now-dissolved-in-rust but formerly intricately shaped brake shoes, the shriveled leather cone of the clutch—all of these and a seemingly endless list of other items are something else again.

Fortunately, the restorer's path has been trodden before, and there are suppliers for these items, or craftsmen who will make what cannot be supplied.

As the Packard (let us say) comes apart, the logic of its assembly becomes apparent. Doors were made of beautifully joined wood of high quality, its joints mortised and tenoned, sanded, and varnished even though designed to be covered forever by the metal doors. The doors themselves, from the inside out, are seen to be masterpieces of metalwork, joined with welds that were smoothed and filed.

Each part that went into the car received the attention of a specialist who performed the task over and over with specially designed tools. The restorer ingenuously approaches each of these specialists' task with a fresh slate: no experience, no knowledge, and no tools.

Neither the work nor the parts are inexpensive.

Dust-covered, square-bodied cars form a suitably drab backdrop for the luscious curves of the 1938 Delage D/8-120.

Plate glass for the windows, for example, can be acquired at about $11 per square foot. A front fender might cost $500, a glove compartment door $75, a distributor $275 (they're pretty rare), and a replacement engine $5,000.

The restorer is sure to encounter peaks and valleys over the months (years) of work. The peaks come when a bit of work is done perfectly and gets a smile of approval from someone knowledgeable; the valleys occur when weeks of work must be scrapped because of a previously unseen flaw. But eventually it becomes evident that virtually all the parts have been located, and the vehicle starts coming together in recognizable form. Interiors are brought back to showroom newness with original fabric made by the same mill out of the same raw materials, and the refinished dashboard gets its shining set of restored instruments.

As the project nears completion, the disquieting thought now arises that the interior of the purged, cleansed, and redone gasoline tank must be stained with fuel; the crankcase of the engine must be sullied with oil; and water must be placed in the gleaming radiator. Ultimately, the engine must be set up, timed, and otherwise prepared for actual starting and running.

Finally, however, the day comes when the roll-out is made, when the car is complete, when the engine has been run and the brakes checked and the clutch tested. Then the restorer gets behind the wheel, proud of the efforts that made this moment possible, and conscious of many things. Most striking is the realization that the time and perhaps $20,000 heretofore invested have resulted in a Class 2 condition car—one that could, with more expertise, be brought someday to Class 1. And given that a Class 2 car could command as much as $35,000 and a Class 1 car $50,000, the restorer has recovered the initial cash outlay for hulk and parts, and has earned as much as 20¢ per hour for labor. Not that a proud restorer would ever sell, of course—not unless chance intervened during another drive in the country, and there were another farmer with another hulk of a rusting classic.

C H A P T E R 7

FOREWORD TO THE FUTURE

EVANESCENT DREAM CARS

The automobile industry has made a habit over the years of fearlessly predicting the future, almost always incorrectly. For years the moguls of the auto industry made beguiling predictions of exotically shaped cars, powered (perhaps) by atomic energy and driven on automatically controlled highways. In doing so they utterly failed to anticipate the devastating fuel shortage of 1973, and now they are busily engaged in ignoring the next one. As a group, American car makers looked with disdain on Volkswagen and contempt on Toyota, while fecklessly building muscle cars.

The predictions about future dream cars were laden with the basic error of the industry: that consumers would be satisfied with ''advances'' in styling, and not demand (or respond to) improvements in utility or economy. The result was often performance-unrelated but beautiful body work, elements of which eventually did appear in production cars. The dream cars themselves were usually just that, evanescent dreams from which only a fender line, a windshield, or a name would persist.

Perhaps the dream cars' greatest contribution to the industry was in their reverse validation of the venerable principle that form follows function. As the dream cars were without function, they could have any form. Pointlessly long hoods, swooping fenders, skirted front wheels, and improbable driving compartments were carelessly conferred upon cars that would never travel a distance greater than the radius of a turning platform at an automobile show.

It didn't have to be this way. The dream cars created by Raymond Loewy's industrial design team were intended for production and were models of beautiful form following economical function. Many of Loewy's Studebaker designs—from the 1939 Champion through the 1941 President Skyway to the famous ''going or coming'' postwar models—would be categorized as classic if they had not been mass-produced. If there had been only 3,000 1947 Studebaker Commander Land Cruisers, or the same number of 1953 Commander Regal Starliner Coupes built, they would be as hotly desired today as Cords or Auburns. The later Loewy Avanti illustrates this. Built in small numbers on an inexpensive Lark frame, it became a true classic that was retained in production for more than twenty years. (While Raymond Loewy is always given the credit, his team included such great designers as Bob Bourke, Virgil Exner, and—later—Holdon Koto and Robert Andrews.)

Yet the dream cars were fun, and perhaps that

LEFT:
In the fifties, GM was still the King of Styling and a "dream car," like this 1957 LeSabre at the Paris Auto Show, would attract capacity crowds. No one seemed to mind that few, if any, of the mechanical features would appear in production cars. It was enough if elements of the styling—or even just the name—survived to grace cars that could be purchased.

RIGHT:
The jet aircraft had as much influence on post-World War II styling as the airplane itself had on automobiles after World War I. The Pontiac Firebird II was a roadworthy four seater that used a number of advanced ideas, including a heat exchanger to recoup energy from the exhaust, titanium body parts, an electromagnetically controlled four-speed transmission, and self-leveling hydropneumatic suspension. Of all of these, only the Firebird name survived the transition into production.

OVERLEAF:
A brilliant photographic idea: Ford's concept car—the 1987 Brezza—posed in front of square-crunched predecessors.

was enough to justify their existence. The earliest ones were the most influential, while later ones served more as props for leggy auto-show girls than as showcases for innovative engineering hardware.

Most people consider the first "official dream car" to be the handsome Y-job designed by Harley Earl in 1939. It was basically a standard Buick with a lean-lined body, and it did in fact influence Buick's exterior design through the early 1950s.

Prior to the Y-job, other specially designed cars were sometimes produced as startling styling exercises on drivable chassis. Among these were the sleek Pierce Silver Arrow of 1933, the Cadillac Aerodynamic coupe of the same year, the lovely Cord-like modified Buick that appeared in the Topper movies, and many products from European coachworks.

Chrysler followed GM's example in 1940, creating the Chrysler Newport and Thunderbolt show cars. While both were amply fleshed in flowing sheet metal, they retained elements of traditional styling. The Newport was in fact a topless dual-cowl phaeton, while the rotund Thunderbolt could be mistaken from the side or rear today as a postwar Packard. Ralph Roberts did the Newport, while Alex Tremulis (most noted for the later Tucker) did the Thunderbolt. Both cars imparted lines and motifs to Chryslers until the 1950s.

When the automotive sellers' market began to decline a few years after the end of World War II, General Motors began designing a series of dream cars that grew progressively more radical as the number of ideas that were nontransferable to ordinary use increased. Ford and Chrysler generally followed suit. The familiar interpretation of aviation motifs began to be supplemented by space-age inspiration, and where propeller hubs had been adequate for the postwar Studebaker updates, jets and rockets came to symbolize the next era of dream cars.

The 1951 Buick XP-300 was an attractive two-

EAGLE GT
GOODYEAR

seater which differed from most subsequent dream cars in its advanced chassis and engine features. Its layout hinted at a turbine engine, but the power was supplied by a supercharged V-8. The XP-300 featured a "jet exhaust" but lacked a "jet intake," which was included on the LeSabre of the same year. It also had fins, flowing lines, and garish amounts of chrome, but it lacked the LeSabre's two deadly Dagmars in the bumper.

Chrysler wanted to enter the arena again and turned to Italy for styling help. The best of the joint products was the 1951 Ghia Chrysler K-310, an attractive five-passenger sports coupe. Dodge in 1953 had the firm execute a series of Firearrows that were excellent designs. Unfortunately the only effect they had was on the naming of Dodge cars of the period, and not on their design.

Chrysler was lucky. The Italians, who created such beautiful cars for themselves, often seemed bent on proving that American car manufacturers had absolutely no ability to recognize a tasteless design. When United States management hired them to put Detroit cars in Milan suits, they responded with ugly travesties that could scarcely have been presented with a straight face. Pininfarina should have been ashamed at what he did to the bulbous Nash line, and Carrozzeria Touring's Hudson Italia was a harsh practical joke on an undeserving company. The Italia was a textbook example of committee styling, and the result injured the reputations of all.

During the second half of the century, American production cars entered a period of extravagant excess that was exaggerated in the dream cars. The human form, for years squeezed by designers into spaces suitable only for caper-stuffed anchovies, was now deemed temperature-proof. Cars came with huge transparent roofs guaranteed to fry-cook drivers faster than a Big Mac. Doors popped up, folded over, or recessed into hidden compartments, doing everything but make entrance easy. Seats swiveled, reclined, and ejected. Instruments glowed in dazzling computer colors, pretending to control routes and arrivals automatically. Fins approached the proportions later seen in *Jaws,* while jutting fenders, spear-like hood ornaments, and copious exhausts bellicosely threatened pedestrians.

There are still dream cars, in less extravagant form, but the audience has become jaded. At recent automobile shows, the relatively tame cars of the future being presented simply have not commanded the attention of press and public. It is difficult, after all, to compete with exotic production cars that stretch the imagination at every viewing. One look at the finest contemporary classic, the Aston Martin Lagonda or at the outstanding (though no longer new) sportscar, the Lamborghini Countach, proves that dreams are already on the road.

OPPOSITE:
The Chrysler Thunderbolt of 1941 *(top)* featured the softly rounded bulbous envelope styling which would characterize many non-GM cars after the war. A push-button freak's dream, the retractable top, headlights, windows, and door locks could be controlled with a tap of the finger. Body was by LeBaron.

In 1955, Lincoln Mercury designers sent their ideas to Italy to be built by Ghia. The result was this double-bubble–Lincoln Futura *(bottom)*, which managed to maintain the flavor of the good-looking contemporary Lincolns while adding fins as well as a double-sliding roof that was equally handsome —but totally impractical.

Future Engineering

The same driving intensity that has resulted in the current crop of remarkably good automobiles is working for the future. Exciting new developments in the works will bring the automobile to an even higher state of refinement. Later in this chapter we will consider whether the developments are really necessary.

Predictably, most of these future improvements are not tied to extraordinary styling changes. Instead they involve an immensely clever application of computer capabilities and new materials to engines, chassis, and running gear. Let's look at the most significant.

The conventional piston engine, augmented sometimes by turbos and refined by continual computer diagnosis of all of the conditions necessary to run at peak efficiency, will probably not be changed significantly in the next decade. Continued emphasis will be placed on economy, emissions control, and power—seemingly mutually exclusive characteristics that modern engineering has managed to combine with far greater success than could reasonably have been hoped for. The Wankel rotary engine that Mazda has used with great success will also continue to be produced, with similar improvements.

A lot of work is being done on ceramic engines,

which offer lighter weight, better heat exchange, and other advantages. A ceramic turbo is being run in a Nissan 300Z; and the Polimotor firm in Fair Lawn, New Jersey, has created a 2.0-litre dual-overhead-cam, fuel-injected, plastic engine that develops 318 horsepower at 9,500 rpm and weighs about half as much as an ordinary steel engine weighs. The 9,500 rpm is the clue to performance, because that figure is possible only with extremely lightweight moving parts. The Polimotor has run in a Lotus chassis and done well. Extensive use of plastic engines in production cars undoubtedly lies in the distant future, although we can expect to see more ceramic components in the near term.

The on-board computer, which has improved engine operation already, is capable of tuning the chassis and suspension to meet constantly changing road conditions. Manufacturers have sought various ways to control suspension—from early efforts that snubbed shock absorbers to air- and hydraulic-operated suspensions. In all of them, however, the almost infinite number of potential combinations of speed, road, weather, and driver input conditions often dictated that the suspension first follow and then operate in opposition to the cycle of driving events.

The computer, in concert with a mechanical system designed to use it, interprets existing conditions and anticipates future ones, based on information it receives from multiple sources. Just as with the engine, however, if you lose your computer, you lose your control.

FOUR-WHEEL DRIVE

The concept of driving all four wheels for heavy vehicles was first investigated by Ferdinand Porsche at the turn of the century. In the past decade, four-wheel-drive cars for off-road use have become increasingly popular. The term *all-wheel drive* is preferred by Audi engineers, who developed the Quattro as a high-performance luxury sedan in 1980. The company has steadily refined the system since then, and has been followed into the arena by Honda, Subaru, Volkswagen, and others.

The proponents of four-wheel drive claim that it has numerous advantages: power is delivered more evenly to each of the wheels; less slippage occurs; and improved handling and traction are evident on all-weather roads. Road tests of almost identical Audis—one with four-wheel drive, and one without—suggested that four-wheel drive definitely improved performance in adverse weather conditions, but that it made little difference on dry pavement.

Yet four-wheel drive has sales appeal, and Audi and others that offer the system make their pitch in computer-speak advertisements tailored to the techno-gadget crowd.

And if four-wheel drive is good, shouldn't someone be investigating four-wheel steering? Of course, and it's on the way. Mazda's handsome 1986 RX-7 introduced a dynamic tracking suspension system that allows the rear wheels to respond passively to forces introduced in a turn. The result is better handling, less opportunity to skid and slide, and a wonderful selling point.

The logical follow-up to Mazda's passive system will be a fully computer-controlled system that operates in such a way that you can pull up to a parking spot 1 inch longer than the length of your car, make a 90° turn of the wheels (front and back), and simply slide in. Ultimately, we will be able to do the same thing at high speed

The 1988 Aston Martin Lagonda achieved what Jaguar sought, and failed, to do: improve an unbeatable classic. Neither car was out of date in absolute terms—only in the relative sense that the designs had been around for too many years to accommodate marketing goals. Lagonda's subtle refinements (if six headlights can be called subtle) made the enormous sticker price even more attractive. Jaguar, however, merely neutered its classic design, making it bland rather than distinctive.

on the freeway—a prospect that opens whole new vistas in fist-shaking interpersonal relations, and in long-term financial planning for body-and-fender garages.

Less modern than four-wheel steering, the modern antilock braking system found in Audi, Mercedes, and other luxury cars operates via a computer to allow drivers to stand on the brakes, and let the system adjust to avoid wheel slip and skids. Stopping distances are not greatly reduced from those possible with manual braking, but the system avoids the catastrophic loss of control that improper braking can cause.

WHAT DOES IT ALL MEAN?

The real import of these improvements can be quantified. A car with four-wheel drive, four-wheel steering, a turbo-charger, and an antilock braking system can probably go through an automobile test course 5 seconds faster than a competitor without these items.

Think about it: in the next decade, perhaps as many as 1,000 people will actually take a car through an automobile test course. That translates into 5,000 seconds saved, for an investment of only a few billion dollars!

It may be protested that all of these systems can help save human lives, and this is true; but far more lives could be saved by requiring everyone to drive 10 miles per hour slower. What we are really being asked to focus on is the ephemeral pleasure of the very few people who can tell the difference between driving at 110 mph on a winding road and driving at 100. The vast majority of drivers rarely exceed 65 mph on the straightaway and prudently throttle down when the traffic signs tell them to do so.

So as disquieting as it may be to advertising agencies and advocate engineers, the effort likely to be (and already being) devoted to almost irrelevant refinement should be focused instead on developing radically new means of transportation that will be not just desirable, but mandatory for the twenty-first century.

The problem is identical to the one that characterized the American car industry from 1950 to 1975. There is no short-term incentive for manufacturers to change their practice of building ever-better, ever-costlier cars, even when the same manufacturers employ economists who can plot to the year when the next (and terminal) fuel shortage will occur.

THE LOWDOWN

Of all the improvements presently under development, the antilock braking system is the most important to safety and utility. The sad fact is that the billions of dollars, marks, lire, yen, francs, and pounds being spent on sophisticated engines, all-wheel drive, four-wheel steering, and the rest are terribly misdirected. These developments should be left to builders of racing cars, while the industry giants face up to the genuine problems that threaten them with extinction.

Nothing can avert a serious petroleum fuel shortage in twenty years, and a catastrophic petroleum fuel famine in fifty years. Even if exploration for new petroleum fields were accelerated, conversion of coal to engine fuel were resumed, and a 100 percent improvement were made in the average fuel economy of cars, we would still face a long-term insurmountable obstacle to driving as we know it today: all of our fossil fuels are being used up, and we can no longer sustain wildly extravagant automotive habits. Obviously, we should do something about it, and we can.

THE CENTURION:
CAR OF THE TWENTY-FIRST CENTURY

We is the operative word, of course: we collectively as consumer, car buff, politician, statesman, voter, and human being. Let *us* design a totally new car, and let us begin by taking a fresh look at the concept of the automobile as transportation, not romance. We'll call our car the CENTURION, because we still have to have a good name and

Most people assumed that the good-looking open car in *Topper* was a modified Cord—thanks to its squarish bonnett, supercharger pipes, and pontoon fenders—but it was actually a Buick. That's Cary Grant and Constance Bennett pretending slumber before an appreciative crowd of extras.

because the term anticipates the approaching period when the fossil fuel shortage will not be an OPEC irritant but an irreversible fact of life. Even as you read, car production is booming around the world, each individual vehicle heightening demand for the diminishing pool of resources and making its contribution to traffic congestion and air pollution.

The next time you are on the freeway, take a quick twenty-car count to see how many of them contain more than two people. You'll probably find that twelve to fifteen of the cars have only one person inside, and only one or two of them have more than two occupants. Yet we implacably build most cars with seats for four or five. Why do we do this? Because manufacturers have found that building a four-seat car with 150 horsepower costs very little more than building a two-seat car with 50 horsepower, and that people overwhelmingly prefer to buy the former, given the choice.

The next time you are within city limits, run a careful average of your driving speed. Almost certainly it will be less than 25 mph if the average is taken any time during the working day. Why then do we demand cars capable of cruising all day at 95 mph?

The next time you take out your computer, try estimating the real cost of driving your car, including depreciation, insurance, fuel, oil, repairs, maintenance, licensing fees, and all other expenses. The numbers are frightening. We complain about the cost of steak, hospital rooms, and college educations while manfully (or womanfully) paying 50¢ to 75¢ per mile for the pleasure of driving.

There *is* a better way, but it requires that we completely revise our thinking. The payoff—the

CENTURION—will justify our efforts, however, by transporting us in style, comfort, and economy. Let's consider the task before us in two steps.

STEP ONE: THE NEW APPROACH

The most difficult thing we must do is to demythologize the car as romantic conveyor of all human emotions. A car is an object for transporting people and cargo from one point to another. We don't take supersonic transports from Washington to Philadelphia because doing so would be grossly inefficient. We shouldn't drive a conventional car 32 km/20 miles back and forth to work through traffic with one person on board, for the same reason: gross inefficiency. The CENTURION will be designed with efficiency uppermost in our minds.

The next most difficult job facing us is to obtain genuine improvement in public transportation. For this, fortunately, we do *not* need to take a survey, install a computer, or hire a consulting firm. Instead, we can simply call back the streetcar—that lovely, electric-powered, convenient, nonpolluting, economical people-mover of the past. And we don't need to ask an aerospace firm to make them; we can simply go to the St. Louis Car Company or to any other professional streetcar company and hand it a construction contract. We don't need the streetcars to be magnetically levitized, contained in a vacuum tube, suspended from a monorail, or superconducted. Just two rails on the ground and an electric line above, please, and the problem is on its way to solution.

Interurban transportation can be handled by a combination of present-day conventional automobiles (with appropriate pooling), interurban electric trains, conventional trains, and aircraft.

STEP TWO: DETERMINING THE CENTURION'S CHARACTERISTICS

The beauty of our concept is that we don't have to give up cars; we just have to give up obsolete cars. In return, we can maintain our freedom of mobility and pay a lesser price for it. Right now, a whole series of independent events are occurring that would enable us to achieve a modern car in ten years—probably a far shorter time than it would take to modernize the public transportation system.

The first independent event is the continuing revolution in computers. The present generation of computers is adequate for the task proposed here, but these computers will be superseded every few years by vastly more powerful, lighter, and less expensive computers for on-board use in the CENTURION. Engineers already possess the computer capability to design what we propose.

The second independent event is the wholesale introduction of new materials and new fabricating techniques, stemming from aviation and space applications. The high-strength, lightweight epoxy materials used in such diverse applications as the Rutan Voyager (which flew nonstop and nonrefueled around the world) and the stealth bomber are of enormous potential benefit to the CENTURION. Once they are applied, a virtually crash-proof structure—lightweight, resilient, and easy to repair—will surround the occupants.

The third independent event is the vast continuing improvement of solar and electrical power sources, energy-conversion devices, electric motors, and power transmission systems. The CENTURION can be an immediate beneficiary of these improvements.

The fourth independent event comprises the extraordinary improvements being made in the aerodynamics and the power requirements of human-powered vehicles. Our plan envisages replacing such anomalous activities as jogging and aerobics with auto travel workouts: the CENTURION can provide all of the fitness a person could ask for while saving money by the ton. Human power is only part of the solution, however, so people who are physically disabled or simply uninclined to use muscle power can turn to an alternative source of power.

The intelligent integration of these simultaneous, independent events will give the CENTURION the following characteristics:

1. Passenger Capacity: The car can seat two adults and two children. Variants are available in a number of styles (for example, with two sets of human-power inputs, a stretched car body).

2. Combined Power Sources: The primary power source is electric, supplemented by human power as desired. Solar energy is used as a sustaining source, but the CENTURION is routinely plugged in for charging overnight. The secondary power source is a small two-cylinder gasoline engine, used both to charge the new power storage units and to provide energy directly to the wheels. The tertiary source consists of sophisticated pedal systems lifted directly from current human-powered flight experience; these can be used to add power to the car and to help charge the storage units.

3. Performance: Combined human and electric top speed is 45 mph, with an average cruise speed of 25 mph. The CENTURION's range is 150 miles without use of either solar energy or the gasoline engine. A small computer controls gear shifts.

4. Safety: Because of the car's low weight and low speed, its reinforced carbon-fiber construction renders it almost indestructible. Commuter routes are now restricted by law to cars of a similar type, so the worst case—a head-on collision—is still easily survivable.

5. Cost: Development costs are less than those for four-wheel-drive research during the gasoline era. Unit cost, at production levels comparable to those for late twentieth-century cars, is around £3,450/$6,500. It must be understood that the CENTURION is not a glorified bicycle, but rather a quality automobile intended for 100,000 miles of use. The components are built to quality specifications that match those of the best cars of the past.

6. Human Effort: The primary motive force for the car is electric; the human-power input is selective and optional, but can be used both as a means of economy and as a fitness device.

7. Equipment: The twentieth-century vision of an automobile interior as a combination coach, bedroom, video store, and F-15 cockpit has been dispensed with. The CENTURION's driver and passengers can be kept comfortable by reasonable-sized heating and cooling units and intelligently designed seats. Instrumentation includes the usual speedometer and odometer, supplemented by the special gauges needed for the electric motor and gasoline engine. An optional device is a meter to record the level of human muscle effort in terms of calorie consumption.

8. Appearance: The car of the future is low; seats are semireclining; exterior surfaces contain lots of clear but strong plastic, suitably treated to reflect heat. Wheels are thin carbon-fiber monocoque structures similar to those used on advanced bicycles. The CENTURION's body is aerodynamically refined not only for low frontal drag, but for resistance to cross winds.

9. Appeal: With gasoline priced at $1 per gallon, the CENTURION's appeal would not be great; with gasoline priced at $20 per gallon and a week's wait to get it, the appeal is tremendous. Remember that the choice is not between the CENTURION and older-style cars; it is between the CENTURION and nothing.

Everything comes in its own time, and unfortunately everything goes in its own time, too. The time left before fossil fuels go to the ash heap of history is a matter now of decades, not centuries. The situation calls for a complete revamping of world thinking about automobiles, and the CENTURION is a feasible step in the right direction.

While our discussion of the CENTURION has been perhaps a bit tongue-in-cheek, it is intended to spur serious thought about problems we have ignored for almost half a century. The time for ignoring has just about expired. The next fossil-fuel crisis will leave the world with the mobility of a beached whale. Something has to be done, and the time for us to begin has already passed. It is imperative that we begin now, and make up for lost time. And who knows: if we tried cars of the CENTURION mode, we might even find we liked them.

ENVOI

Despite the change that must take place in cars of the future, those of the past will always be loved for many deeply felt reasons. For more than a century, the automobile has been a major force in Western culture, and its full impact is not yet known.

We have seen how cars have changed over time, and how certain individual cars seem to grip our imagination in ways that are difficult to explain. Not everyone enjoys automobiles; and indeed, some people see them as necessary evils. But for most of us, there is a particular favorite car—one that somehow speaks to the soul of its would-be owner.

And the true car buff has many favorites, depending on background, mood, and experience. In the best of all worlds, where garages grow on trees, insurance isn't necessary, and gasoline is as plentiful as the sea, you could have a stable of cars to suit every need. Let's pretend it is the best of all worlds, and see what cars would provide the greatest satisfaction in various categories. The only limitation (but a difficult one) is to select just three cars in each category.

OPPOSITE:
(Top): **A Duesenberg SJ Torpedo Sedanette with body by Bohman & Schwartz. *(Bottom):* Normally not the best for a relatively small car, the color black seemed to emphasize the low racey lines of the 1937 Cord 812 Berline. It is unfortunate that business problems prohibited the long-term development of this brilliant design.**
OVERLEAF:
The 1964 ½ Ford Mustang Coupe, the instantaneous success that started the "pony car revolution" and propelled Lee Iacocca onto a path that led to the leadership of Ford and Chrysler, best-selling books, and frequent mention as a potential U.S. presidential candidate.

UTILITY CARS

Cars are first and foremost to be used, and we should logically have half a dozen here to meet our varying needs. For short trips, a 1987 Honda CRX is hard to beat. For bigger loads, a 1987 Chevrolet Caprice Classic five-door wagon would be adequate. For travel around town or long-distance trips, a new Mercedes-Benz 300E fills the bill.

CLASSIC ERA

There are some givens for the classic era. A limit of three is almost impossible not to exceed, but even if only three could be chosen, two would certainly be a 1937 supercharged Cord and a Duesenberg Torpedo Phaeton. The appeal of these cars is overwhelming, and they must head the list. The competition for the third position is formidable, but the need for Gallic grace on our list calls for a Bugatti 57C Atlantic.

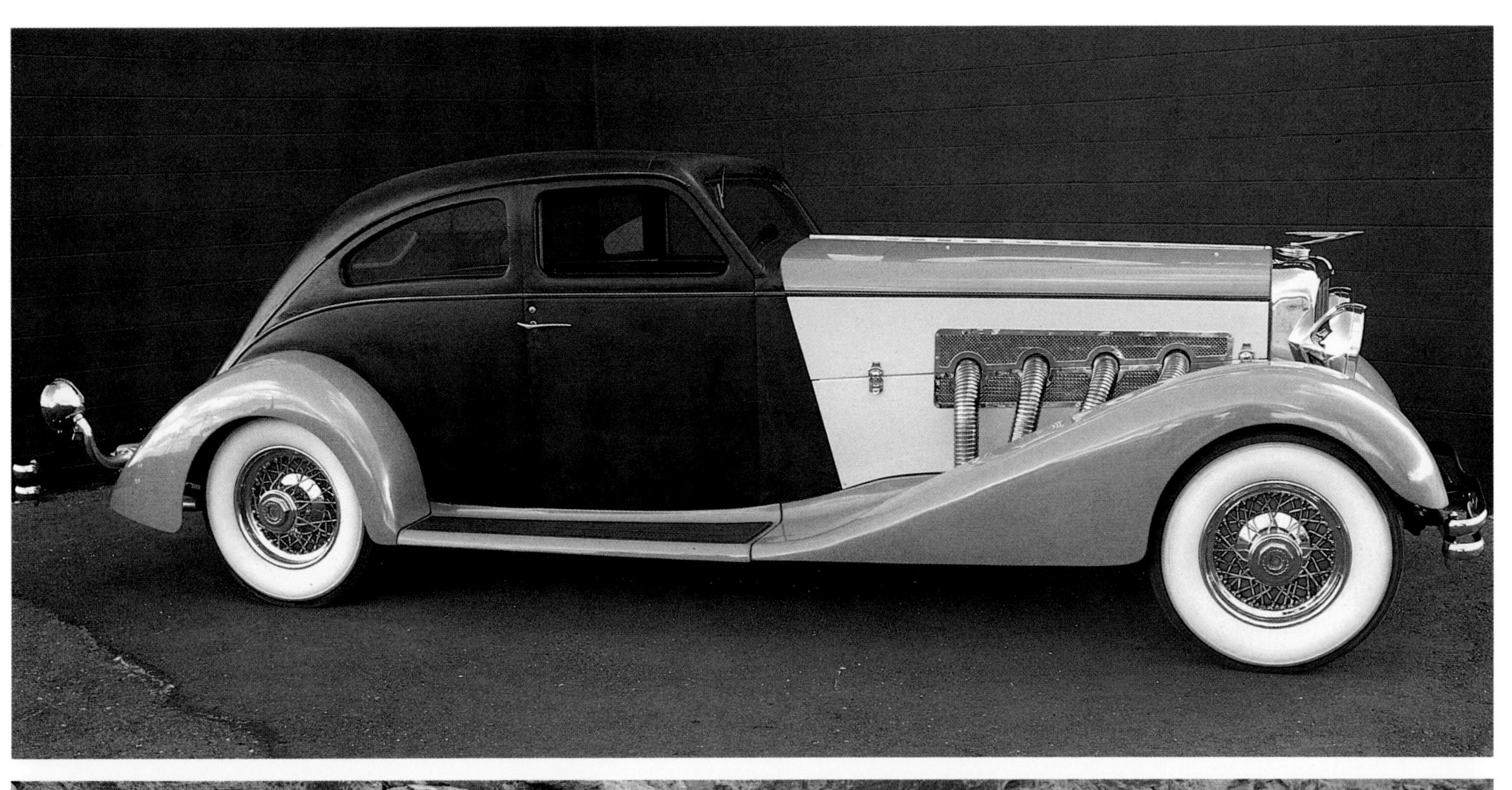

Coca-Cola

Gasoline
Billy's
LUBE
and
OIL
Coca-Cola
IN BOTTLES
DRINK
Coca-Cola
ICE COLD
DRINK
Coca-Cola
ICE COLD

ANTIQUE ERA

The supreme antique would of course be the original Rolls-Royce Silver Ghost, and its stablemate must be a good solid 1927 Model T. A late-model Doble steamer would provide both utility and variety.

SPORTS CARS

This category offers a mixed bag of delights. For today, a Lamborghini Countach would be a marginal choice over a Ferrari Testa Rossa. A long leap into the past would light upon a 1930 Bentley, overstressing blower and all. And the Jaguar XK-120, while perhaps not the greatest of sports cars, was the most influential postwar type, and deserves an honored place in our collection.

BELOW:
The horseshoe grille of the late, now-lamented Edsel was the subject of a thousand one-liners, one of the best being that it looked like an Oldsmobile sucking a lemon. ***(Photograph by Frank S. Balthis)***
OPPOSITE:
The Volkswagen Convertible ***(top)*****. An open-top Bug was just about the cutest car imagineable during the period.**
Whatever fads and economic trends come and go, demand for the Chevrolet Caprice ***(bottom)*** **persists. Some people just want a big, solid car with lots of room, and they are willing to pay whatever extra is required in operating costs (if anything). Not only has this 1987 wagon retained its size, it is also of genuinely better quality than its predecessors.**
PAGES 228–229:
A mule is often considered the classic hybrid, a unique, functional combination of two separate bloodlines; as such, beauty is not a criterion. The Bugatti 57C Atlantic is a hybrid of another sort—one blending art and the automobile; one that achieved the ultimate in beauty as well as performance.
PAGES 230–231:
The Rolls-Royce Silver Ghost—perhaps the most dramatic instance of a single car capturing the soul of an empire. For the world, the Rolls-Royce Silver Ghost was the embodiment of the English ruling class.

MUSCLE CARS

Muscle cars are somewhat daunting and are certainly not everyone's cup of tea; nonetheless they form a part of American culture, and no collection would be complete without the one that started it all, John DeLorean's 1963 Pontiac GTO. The 1968 Plymouth RoadRunner was a much better car in almost all respects than the GTO but, despite its appealing name, never received the popular recognition it deserved. Today, there are a host of muscle cars from which to choose, but a hotted up Ford Mustang, as ancient in design as it is, is probably the best of the bunch.

FUN CARS

They are all fun cars, of course, but to select a final three, quasi-independent of category, let's take a Volkswagen Bug, a 1953 Cadillac El Dorado, and a 1958 Edsel. They represent extremes in utility, styling, and acceptance, but each one will today turn heads when you drive.

Of course, there are many other categories and many other cars to fit in them, but the eighteen listed above offer enough appeal for a lifetime.

Certainly, it would have been a pleasure to include some in-depth discussion of racing cars, record cars, fantasy cars, and other genres not covered; but a close reading of this book should make clear that it is not really about automobiles after all, but about the guiding spirits behind them—designers, builders, buyers, and drivers. And no matter what the category of car and the period during which it was produced, these guiding spirits all share the same fundamentally human characteristics. The automobile has brought out the best in man's creative spirit, even if its use has not always been in man's best interest. At some far-off future time, when human beings are simply levitated from one spot to another, and neither cars nor ships nor airplanes are required, a beautiful car will continue to stir warm, fundamentally human responses in all who see it.

EXK 6

R 1064

PHOTO CREDITS

Pages 1, 3 (both): © Association des Amis de J. H. Lartigue
5: John Vachon/Collection of the Library of Congress
8: The Bettmann Archive
9: © Lucinda Lewis
12: © 1969 James Van Der Zee
13: Culver Pictures
16–17: © 1981 Ruth Orkin
19: © Ernst Haas
20, 21 (both): Culver Pictures
22: Russell Lee/Collection of the Library of Congress
23: © Robert Frank/Courtesy Pace/MacGill Gallery, New York
24–25: Henry Austin Clark, Jr., Collection, Glen Cove, New York
28–29: Courtesy David Diamond Collection
32–33, 36, 37: The Bettmann Archive
40–41: Courtesy Fiat, Turin, Italy
43: Alfred Stieglitz/The Cleveland Museum of Art, Gift of Cary Ross
44–45: © Association des Amis de J. H. Lartigue
49: F.P.G. Int'l.
56: Collection of the Library of Congress
57: Henry Austin Clark, Jr., Collection, Glen Cove, New York
58–59: From the Collections of Henry Ford Museum and Greenfield Village
60–61: © Association des Amis de J. H. Lartigue
61: Collection of the Library of Congress
62–63: Culver Pictures
65, 66–67: Courtesy Maurice de Montfalcon
69 (top): Arthur Rothstein/Collection of Library of Congress
69 (bottom): From the Collections of Henry Ford Museum and Greenfield Village
72, 73 : Henry Austin Clark, Jr., Collection, Glen Cove, New York
76–77: © Association des Amis de J. H. Lartigue
81, 85 (all): © Lucinda Lewis
88: Courtesy Danmarks Tekniske Museum
89: Collection of the Library of Congress
92–93: © Association des Amis de J. H. Lartigue
97, 100: The Bettmann Archive
101: Courtesy Lancia, Turin, Italy
104–105: L'Illustration/Sygma
108: The Bettmann Archive
110: R. Lessmann/The Bettmann Archive
112: The Bettmann Archive
116 (top): F.P.G. Int'l.
116 (bottom): The Bettmann Archive
117: F.P.G. Int'l.
120 (all), 121: © Curtice Taylor
123: © Willy Maywald—ADAGP/PARIS
124: The Bettmann Archive
125: Culver Pictures
129 (both): The Bettmann Archive
132: From the Bostwick-Frohardt Collection owned by KMTV and on permanent loan to Western Heritage Museum, Omaha, Nebraska
133: Culver Pictures
135: F.P.G. Int'l.
138: Courtesy Diario ABC/Prensa Espanola, S.A., Madrid
142: Courtesy National Technical Museum/Prague, Czechoslovakia
148: John Vachon/Collection of Library of Congress
152: The Bettmann Archive
153: Margaret Bourke-White, Life Magazine, © Time Inc.
156: © Association des Amis de J. H. Lartigue

In the open-air comfort of a convertible, a woman and her pet enjoy the colorful spectacle of fall foliage in the Berkshires, far from the European battlefields of 1941. (Photograph by John Collier)

157: The Bettmann Archive
160–161: © Association des Amis de J. H. Lartigue
164: Courtesy Fiat, Turin, Italy
165: Culver Pictures
168: © Curtice Taylor
169: © Harry Callahan/Courtesy Pace/MacGill Gallery, New York
172: Arthur Siegel/Collection of Library of Congress
173: © Willy Maywald—ADAGP/PARIS
176: © Robert Frank/Courtesy Pace/MaçGill Gallery, New York
177: The Kobal Collection
Page 178 (both): Courtesy Nissan Motor Corporation
181: The Bettmann Archive
182 (left): Courtesy Nissan Motor Corporation
182 (right): Courtesy American Honda Motor Co., Inc.
183 (left): Courtesy Buick, a division of GM Corp.
183 (right), 184 (left): Courtesy Ford Motor Company
184 (right): Courtesy Toyota Motor Corporation
187: The Kobal Collection
189: Courtesy National Automotive History Collection of the Detroit Public Library
190: Courtesy the William F. Harrah Automobile Museum
191: The Bettmann Archive
192: Culver Pictures
193: Courtesy the Collection of Sonia Delaunay
195: Courtesy National Technical Museum/Prague, Czechoslovakia
199: The Kobal Collection
203: © Tim Street-Porter
205: Russell Lee/Collection of the Library of Congress
206: © 1976 Bohumil ''Bob'' Krcil
207: © Curtice Taylor
211 (left): F.P.G. Int'l.
211 (right): The Bettmann Archive
215 (top): Courtesy Chrysler Corporation
215 (bottom): The Bettmann Archive
219: The Kobal Collection
223 (both): © Lucinda Lewis
226: © 1985 Frank S. Balthis
227 (top): The Bettmann Archive
227 (bottom): Courtesy Chevrolet, a division of GM Corp.
230–231: © Lucinda Lewis
233: John Collier/Collection of the Library of Congress

We wish to thank and acknowledge the following for their cooperation in allowing the automobiles in their collections to be photographed for this book:

Edward Aycoth & Co.; Blackhawk Classic Auto Collection; Herb Boyer; Bob and Ellen Cole; The Ford Motor Company; The Henry Ford Museum; Raja Gargour/Hill & Vaughan; James B. Gilliland; Carl Hamon; Jacques Harquindegay; Harrah's Automobile Collection; Russell Head; Gary and Dotty Hein; Heishman-Porsche, Arlington, VA; Jules and Sally Heumann; Hollywood Sports Cars; Burge Hulett; Jim and Penny Hull; Imperial Palace Auto Collection; Peter Mullin; Merle Norman Classic Beauty Collection; Thomas Perkins; Don Prieto; Gail Remy; Louis Sellyei, Jr.; Henry Uihlein II; Leonard, Nancy and Ricky Urlik; Chuck Vickery; Clyde Wade and Sons; Thomas C. Wellnitz; Charles Welsh; Dr. Donald Williams.

INDEX

Pages in *italic* contain illustrations